Transient Liquid Phase Bonding

David J. Fisher

Published by **Materials Research Forum LLC**
Millersville, PA 17551, USA

Published as part of the book series
Materials Research Foundations
Volume 43 (2019)
ISSN 2471-8890 (Print)
ISSN 2471-8904 (Online)

Print ISBN 978-1-64490-004-8
ePDF ISBN 978-1-64490-005-5

This book contains information obtained from authentic and highly regarded sources. Reasonable efforts have been made to publish reliable data and information, but the author and publisher cannot assume responsibility for the validity of all materials or the consequences of their use. The authors and publishers have attempted to trace the copyright holders of all material reproduced in this publication and apologize to copyright holders if permission to publish in this form has not been obtained. If any copyright material has not been acknowledged please write and let us know so we may rectify in any future reprint.

Distributed worldwide by

Materials Research Forum LLC
105 Springdale Lane
Millersville, PA 17551
USA
http://www.mrforum.com

Printed in the United States of America
10 9 8 7 6 5 4 3 2 1

Table of Contents

Materials Research Forum LLC
doi: http://dx.doi.org/10.21741/ 9781644900055

Transient Liquid Phase Bonding

The joining of metals, like the casting of metals, is one of those activities which has long flourished in human society in spite of a complete initial ignorance of the nature of the materials and of the microscopic processes which are involved. It was not until the start of the 20th century that proper scientific investigation of the microstructures of metals began. Within living memory, engineers would still look sagely at fracture surfaces and declare that, "the metal failed because it had crystallized": apparently believing that its normal state was amorphous.

Given this ignorance, one wonders at the longevity of what is here termed: transient liquid phase bonding. Joining two pieces of metal together by melting them both (i.e. welding) would be a rather obvious strategy, but introducing an intermediate layer which might not even have to melt, smacks of incredible sophistication. On the other hand, ice-sculptors have long joined together blocks of ice by sprinkling the surfaces to be joined with salt. So perhaps the mere blind extrapolation of a effective known technique was accidentally found to work. Archaeologists claim that the method was already known some 4500 years ago, on the basis of ornamented grave goods found in Egyptian pyramids[1]. It is also said that Etruscan artists joined gold beads to more massive gold articles by means of so-called granulation, in which an oxide interlayer was used. However, this process was perhaps generally closer to sintering in nature and the interlayer a mere flux rather than an active participant. On the other hand, when copper was present in the interlayer, heating apparently allowed the copper to form a lower melting-point alloy with gold. Solid-state processes could then even-out the local composition and remove any obvious sign that a braze/solder had been used.

Before considering transient liquid phase bonding, it is useful to review more general cases of diffusion bonding. In the case of titanium aluminides, solid-state diffusion bonding and vacuum brazing have been the most widely investigated techniques with regard to producing reliable joints[2]. The transient liquid phase technique has also been called 'eutectic brazing', due to the fact that the interlayers used were specifically chosen so as to form a eutectic with the materials to be joined. Alternative terms have been: 'eutectic bonding', 'liquid interface diffusion', 'solid-liquid interdiffusion bonding', 'diffusion brazing', 'transient insert metal bonding' and 'activated diffusion bonding'. 'Transient liquid phase bonding' itself subsumes variants such as, 'wide-gap transient liquid phase bonding' and 'temperature-gradient transient liquid phase bonding'. In the latter case a temperature gradient is imposed across the interface.

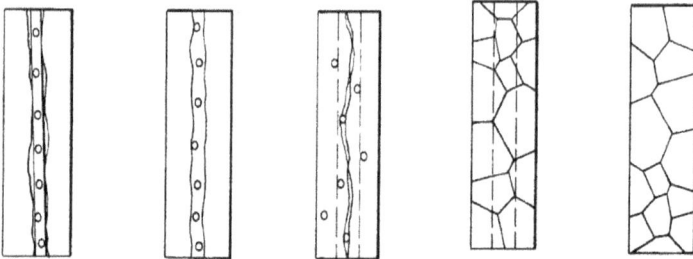

Figure 1. Schematic of the stages involved in the transient liquid phase bonding technique. From left to right: the interlayer is placed between the faying surfaces, the interlayer melts and diffuses into the adjacent solids before solidification completes isothermally.

The interlayer (figure 1) contains a melting point depressant solute and is usually added in the form of a thin foil. The melting point depressant can also be added in a powder form or as a sputtered or thermally sprayed coating. If a simple eutectic is considered, it can be shown that the bulk composition of the interlayer can be tailored to melt at the eutectic temperature, or can shift *in situ* via reaction with the base metal to form liquid. The type-I process involves a pure interlayer while the type-II process involves an interlayer composition which is near to the liquidus composition at the bonding temperature (figure 2). In practice, any composition between these two boundaries may be used in the interlayer but a type-II interlayer is the one which is most often used. This reduces the overall processing-time by reducing the amount of solute to be diffused away from the interface.

Regardless of the name, the basic idea is that the materials to be joined are brought into close contact, with a thin interlayer placed between them. Figures 3 to 6 illustrate schematically the stages through which the joining process then ideally proceeds[3]. As an alternative to a thin foil (rolled crystalline or melt-spun amorphous sheet), a fine loose powder (possibly with a binding agent), a powder compact (sintered or cold isostatically pressed), brazing paste, sputtered deposit or electroplated deposit can be used. It is also possible to 'evaporate' an alloying component out of the material to be joined and create a so-called glazed surface. The most common interlayer thickness is 50μm. It is even possible to leave the interlayer material completely outside of the joint region and let it be drawn in by capillary action. Forces of the order of 10MPa are usually applied in order to keep the parts aligned and to aid bonding but, at the risk of inducing porosity, the parts may also be kept a fixed distance apart.

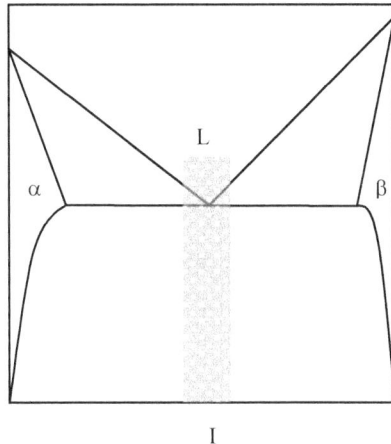

Figure 2. Generic phase diagram illustrating the choice of interlayers. Type-I are pure materials, whereas type-II are close to the eutectic composition

Having set up the bonding arrangement, the whole is heated to a suitable temperature in order to generate liquid in the joint region and is held at that temperature until the liquid has isothermally solidified via diffusion. Heating up to the bonding temperature generally takes from less than one minute, up to about an hour, depending upon the heating method, the heating-rate and the thermal properties. The joint is then homogenized at a suitable annealing temperature. The melting of the interlayer takes from less than one second up to several seconds. The heating is carried out using traditional radiation and conduction methods, or by using radio-frequency induction, ultrasonic, resistance and laser or infra-red techniques. The operation is usually carried out *in vacuo* or in argon. It is rarely performed in nitrogen, hydrogen or air.

If a simple eutectic is used the bulk composition of the interlayer can be chosen so as to melt at the eutectic temperature or can change, once in place, via reaction with the adjacent metal to form a liquid. When the assembly is heated to above the eutectic temperature, the interlayer melts and wets the neighbouring surfaces. The degree of diffusion that occurs depends upon the interdiffusion coefficient between the interlayer and its neighbours, and upon the heating-rate.

Transient Liquid Phase Bonding   Materials Research Forum LLC
Materials Research Foundations **43** (2019)   doi: http://dx.doi.org/10.21741/ 9781644900055

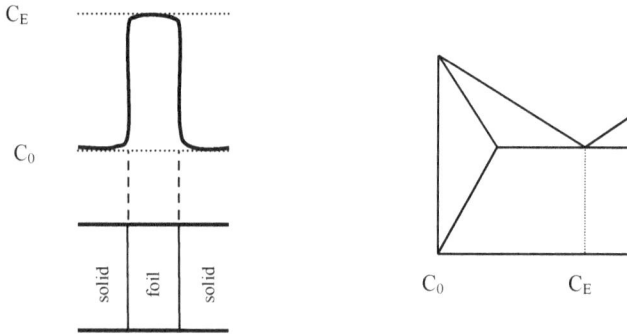

Figure 3. Initial concentration profile of a joint during the assembly stage

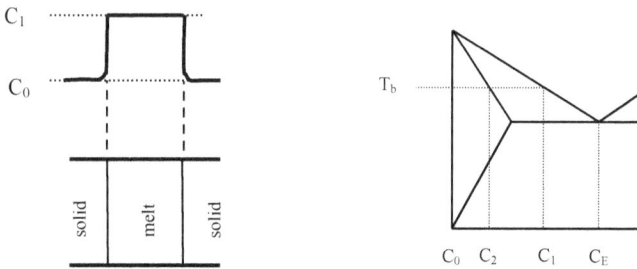

Figure 4. Concentration profile of a joint during melting at the bonding temperature, T_b

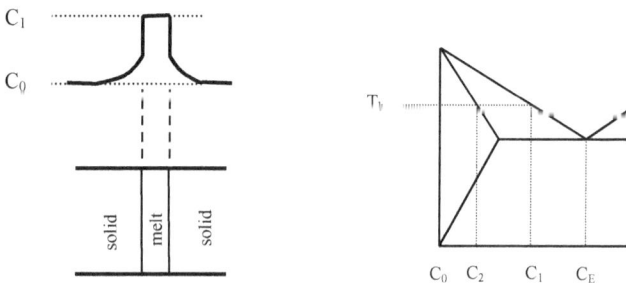

Figure 5. Concentration profile of a joint during the isothermal solidification stage

Because the assembly is essentially a Boltzmann-Matano diffusion couple, recourse can always be had to any relevant equilibrium diagrams in order to predict the course of the process.

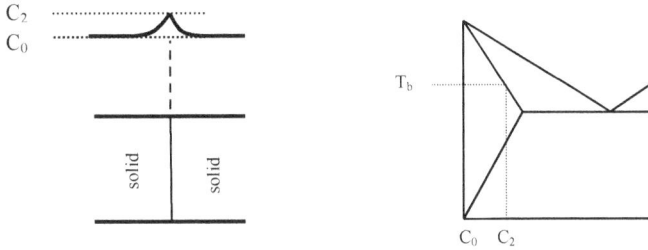

Figure 6. Final concentration profile of a joint

The shape of the so-called mushy zone between the solidus and liquidus lines also affects the kinetics and the optimum bonding temperature, as does the presence of intermetallic phases in the equilibrium diagram. They tend to decrease the diffusion rate. This may be avoided by increasing the temperature (figure 7).

Upon attaining the melting point of the interlayer, the latter liquefies and heating of the joint is continued up to the chosen bonding temperature. This is well above the interlayer melting-point, so as to guarantee complete melting of the interlayer and a high diffusion rate. Reference to a typical equilibrium diagram thus indicates that the liquid composition will be eutectic while the solid composition at the liquid/solid interface will be at the solid-solubility limit. Further heating above the eutectic temperature widens the liquid zone. Increasing the temperature causes the composition of the liquid at the interface to move along the solidus while the solid composition moves along the liquidus (figure 4). The adjacent material is dissolved, and the liquid zone widens; a process which is facilitated by the high diffusivity of the melting-point depressant solute in the liquid. Melt-back takes from seconds to minutes.

The liquid width become greatest at the peak (bonding) temperature. Widening of the liquid can however continue for some time after the bonding temperature is attained, due to kinetic factors. The melt-back distance can be limited by significant diffusion of the interlayer into the substrate before melting and by liquid loss due to wetting of the substrate sides or to the squeezing-out of liquid due to the use of too high a bonding pressure.

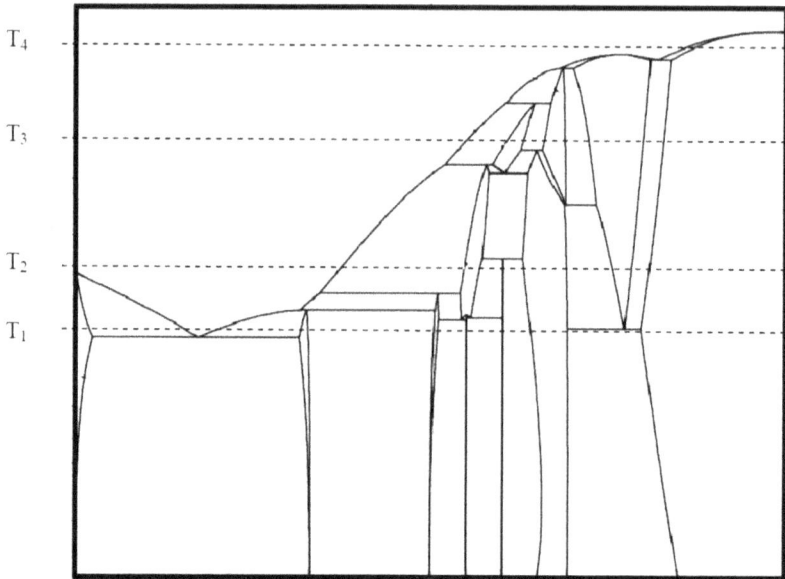

Figure 7. A more complicated generic equilibrium diagram. The use of T_1 would implicate many intermetallic phases, whereas the use of T_2 or T_3 would reduce the number involved and T_4 will eliminate that problem entirely

Too much melt-back of the substrate by the liquid interlayer can have detrimental effects on the final bond in addition to lengthening the isothermal solidification time. And, in some systems, melt-back can reach five to fifteen times the original interlayer thickness. To prevent drastic melt-back that can adversely affect the microstructure, the interlayer should be thin, of a eutectic composition, or of a composition similar to the substrate material.

The above is the idealised situation. There can be complications. For instance, although the much more rapid diffusivity of solute in the melt generally guarantees that the above scenario will occur, it can be envisaged that heating the assembly too slowly might allow the melting-point depressing solute to diffuse into the adjacent solids instead, thus leading to an insufficient liquid volume. It is also not essential that the system be a eutectic one. It is necessary only that the metals to be joined form a low melting-point phase with the solute.

From the macroscopic point-of-view, transient liquid phase bonding consists of the stages: heating, dissolution (widening), isothermal solidification and homogenization. Transient liquid phase bonding is distinct from other brazing processes, precisely due to the re-solidification of the liquid at a constant temperature. The interlayer, rich in a melting-point depressant will, upon heating through the eutectic temperature react with the adjacent metal to form a liquid. During isothermal holding at above the melting temperature of the interlayer, the melting-point depressing solute becomes depleted in the liquid phase via diffusion into the metals to be joined. The resultant solid/liquid interface movement is called isothermal solidification. The desired homogeneous bond appears when isothermal solidification is complete and the advancing solid/liquid interfaces have met at the centre.

Isothermal solidification generally takes from minutes to hours, but there can be outlier cases where it takes from less than a minute or more than 24h. Following transient liquid phase bonding, the melting-point of the bond is similar to that of the joined metals. This feature of transient liquid phase bonding makes it useful for the repair of high-temperature turbine superalloys for example. During traditional brazing, solidification is induced by cooling and thus gives an heterogeneous bond. Isothermal solidification essentially involves epitaxial growth of the joined metals and, after homogenization annealing, the joint will have a microstructure similar to that of the adjacent material. The isothermal solidification mechanism also makes transient liquid phase bonding an attractive method for joining composites and shape memory alloys.

Homogenization can take from minutes to days. A useful rule-of-thumb is that initial melting of the interlayer is an order-of-magnitude faster than the melt-back of the pieces to be joined. The isothermal solidification stage is usually the bottleneck in production processes, given that the homogenization stage takes longer – in principle – but is rarely allowed to go to completion. The process may finally end during the service life of the bonded article.

The overall kinetics also depend upon the nature of the interlayer. Transient liquid phase bonding takes much longer for a eutectic system when pure elements are used. The isothermal solidification stage may be lengthened by a factor of 25 upon using pure elements rather than a eutectic composition. This is because the interlayer has to undergo solid-state interdiffusion with its neighbours at the bonding temperature before liquid can appear. Melt-back is therefore greater.

A combination of high diffusion rates, slow heating and thin interlayer can, on rare occasions, allow all of the interlayer to diffuse into its neighbours before the interlayer's melting-point is attained. In order to avoid this possibility, the interlayer thickness must

Materials Research Forum LLC

doi: http://dx.doi.org/10.21741/ 9781644900055

be greater than a minimum value. Another useful rule-of-thumb is that the length of the isothermal solidification stage is proportional to the square of the interlayer thickness. The interlayer should thus be made somewhat thicker than the critical thickness in order to optimise production times. The processing conditions can also he hemmed-in by other factors. For example, too high a bonding temperature might lead to microstructural breakdown of the pieces to be joined.

In summary, the greatest advantage of transient liquid phase bonding is that the resultant joint can, during service, operate at, or above, the bonding temperature. The joints also tend to possess the same microstructure and mechanical properties as those of the bonded materials. They in fact commonly satisfy the 'golden rule' for a joint: that it should be as strong or stronger than the adjacent material, so that failure occurs elsewhere. Of particular interest to the production engineer are the facts that the process is not greatly impaired by the presence of surface oxide layer, that constraining forces in jigs are much lower than those required by other joining methods, that many joints can be fabricated at the same time and that excess liquid evens-out mating surfaces and thus minimises the need for finishing steps. It can be used when conventional brazing and welding fail, as when the material is susceptible to hot-cracking. The materials which have been successfully bonded include aluminium, cobalt, iron (including stainless steels), nickel and titanium alloys, metal-matrix composites, oxide dispersion-strengthened alloys, ceramics, monocrystals and intermetallics. In this last case, wettability and microstructural development are important and difficulties can arise when bonding intermetallics; although those can be overcome to various degrees[4]. The method also works with 'special geometries' such as cellular structures and micro-circuitry.

A distinction can be drawn between cases where isothermal solidification governs the final microstructure and cases where changes which occur following the completion of isothermal solidification predominate[5]. Practical limitations exist concerning the production of parent-like microstructures and microstructural development in materials where dissimilar materials are involved.

The drawbacks for the process engineer include the facts that lt can be time consuming and expensive. Indeed, the main problem with the method is that, although transient liquid phase bonding can be a reliable method once the processing parameters are determined, the establishment of those parameters can be very tedious ... requiring much trial-and-error metallurgical examination of polished cross-sections of the joints. A promising method is to use differential scanning calorimetry to monitor the progress of such processes.

The basic method can be further 'tweaked' by imposing a temperature gradient. This process leads to an irregular bond interface and usually greater strength. By using the basic method, but filling the original gap with a variety of melting and non-melting layers, the isothermal solidification stage can be speeded-up. Use of a multi-component interlayer can permit a ceramic to be joined to a metal.

Bonding with a temperature gradient imposed across the joint results in shear strengths approaching those of the parent material. Unlike the usual process, where solute diffusion in the solid phase is assumed to be required for solidification to occur, the gradient method relies on solute diffusion in the liquid phase[6].

The width of the liquid phase after the isothermal hold period can be estimated only by measurement of the solidified phase. The initial thickness of the interlayer not surprisingly has a marked effect upon the kinetics of transient liquid phase bonding, as indicated above. Moving-boundary problems in solidification are notoriously difficult to solve mathematically even for a simple eutectic system. In systems where intermetallic compounds can form, analytical models may be very inaccurate. When relevant phase diagrams and diffusion data are all available for the materials used, the isothermal solidification time can be theoretically deduced as a function of temperature. The predictions are very system-dependent; a result of the interaction of the diffusion-rates and phase-diagram geometry. With increasing temperature, the diffusion-rate increases exponentially according to the familiar Arrhenius relationship.

This overall relationship is nevertheless usually parabolic and predicts a minimum isothermal solidification for a given temperature which, in addition, has to lie between the melting points of the interlayer and of the materials to be bonded. In some 'pathological' cases, the system properties can lead to the prediction of monotonically increasing or decreasing times. In the former case the optimum bonding temperature has to be set just above the melting point of the interlayer. In the latter case it has to be set as high as the materials to be bonded can support.

A parametric study has been made of solute redistribution during transient liquid phase bonding, taking account of macroscopic solute diffusion in the liquid and solid phases, as well as solid transformation to liquid due to solute macrosegregation. The overall behaviour could be described in terms of the ratio of solute diffusivities in the liquid and solid, the holding temperature, a parameter which was related to the solidus and liquidus slopes, and the re-melting and re-solidification times (table 1). It was shown numerically that the holding time, the holding temperature and the solute diffusivity ratio strongly affected the solute distribution[7]. This in turn then markedly affected the liquid zone and mushy zone thicknesses. The overall results indicated that the optimum parameters were a high holding-temperature, a long holding-time and a large ratio of the liquidus and solidus slopes of the interlayer material.

Table 1. Dimensionless solute concentration, C/C_{l0}, intervals for the various phase-regions of any alloy as a function of the dimensionless holding temperature, T/m_lC_0, where C_{l0} is the initial concentration of the interlayer, m_l is the liquidus slope, C_0 is the solute concentration, and D_s and D_l are the solid and liquid diffusivities.

T/m_lC_0	D_l/D_s	Zone	C/C_{l0}
0.9755	35	liquid	>0.9756
0.9755	35	mushy	0.1836 to 0.9756
0.9755	35	solid	<0.1836
0.9755	46	liquid	>0.5862
0.9755	46	mushy	0.2522 to 0.5862
0.9755	46	solid	<0.2533
0.9755	56	liquid	>0.9945
0.9755	56	mushy	0.1955 to 0.9945
0.9755	56	solid	<0.1955
0.6600	35	liquid	>0.6600
0.6600	35	mushy	0.1242 to 0.6600
0.6600	35	solid	<0.1242
0.6600	46	liquid	>0.1713
0.6600	46	mushy	0.1713 to 0.3964
0.6600	46	solid	<0.1713
0.6600	56	liquid	>0.6725
0.6600	56	mushy	0.1320 to 0.6725
0.6600	56	solid	<0.1320
0.3442	35	liquid	>0.3444
0.3442	35	mushy	0.0647 to 0.3440

0.3442	35	solid	<0.0647
0.3442	46	liquid	>0.2069
0.3442	46	mushy	0.2069 to 0.0893
0.3442	46	solid	<0.0893
0.3442	56	liquid	>0.3510
0.3442	56	mushy	0.0690 to 0.3510
0.3442	56	solid	<0.0690
0.0290	35	liquid	>0.0284
0.0290	35	mushy	0.0053 to 0.0171
0.0290	35	solid	<0.0053
0.0290	46	liquid	>0.0171
0.0290	46	mushy	0.0073 to 0.0171
0.0290	46	solid	<0.0073
0.0290	56	liquid	0.0290
0.0290	56	mushy	0.0055 to 0.0290
0.0290	56	solid	<0.0055

The present process, in which a liquid layer forms and then solidifies, involves the analysis of moving interfaces. Fixed-grid methods, first developed for simulating temperature fields during melting or solidification have been adapted to the simulation of diffusion-controlled dissolution and solidification[8]. A diffusion equation for the various phases and moving interfaces has been developed by using implicit time-integration.

A so-called supercooled process for bonding has been proposed in which a liquid interlayer of amorphous alloy is treated at a high temperature for a few seconds and then at a low temperature for a few minutes; both temperatures being higher than the melting point. Due to the fall in temperature, supercooling of the composition at the liquid|solid interface occurs and the interfacial stability of the solidification front breaks down; producing a cellular interface. Following solidification, the interface disappears to leave a seamless joint. The method was used to bond carbon steel, using amorphous Fe-Ni as an interlayer. As expected, a non-planar interface formed during the first stage and disappeared during the final stage, leaving an homogeneous joint with no interface[9]. Freed from the detrimental effect of an interface, the impact toughness of the joint was improved to the level of the original material.

The transient liquid-phase bonding of binary alloys can be characterised in terms of two dimensionless parameters. Analytical formulae for the rate at which the liquid region solidifies are valid only in some cases. Numerical modelling produces maps which determine whether a semi-infinite solution provides an acceptable approximation for the analysis of any given system. Further consideration of phase-diagram characteristics, diffusivities and system dimensions could be used to identify the optimum melting-point depressant element to be used in a given situation[10]. It is notable that, due to higher solubilities, those elements which are substitutional solutes in the parent materials frequently permit faster solidification than do those which are interstitial solutes; regardless of the fact that the latter normally have much higher diffusivities.

Successful processing requires an understanding of transient liquid layer behaviour in terms of diffusion-controlled phase boundary migration and capillarity-driven flow. This has been modelled numerically in order to simulate diffusion-controlled dissolution and solidification in one dimension[11]. The model was validated using experimental data from various systems, including solid-solid diffusion couples.

The process of isothermal solidification during transient liquid-phase bonding in a ternary system has been analyzed. Unlike the binary situation, the liquid composition has to change constantly during solidification. When the solubilities or diffusion coefficients of the two solutes are very different, solidification divides into two parabolic regimes; one dominated by the faster solute and the other by the slower solute[12]. In some cases, complete solidification may not be possible within experimentally reasonable times.

A further refinement is partial transient liquid-phase bonding. This is a variant which is normally used to join ceramics and requires the use of a multiple-component interlayer. A typical interlayer consists of a thick refractory layer sandwiched between thinner layers of lower melting-point. There may even be several thin layers on each side of the refractory. The method is related to the wide-gap and active transient bonding methods which were alluded to above, and pre-dates them by some years. Upon heating to the bonding temperature, a melt forms and the liquid wets each of the ceramics to be bonded while simultaneously diffusing into the solid refractory core. The liquid regions then solidify isothermally, as in transient liquid phase bonding, and overall homogenization of the entire assembly then leads to a refractory bond. The refractory core is a foil that can be between 20 and 1000μm thick, and can be made of cobalt, copper, gold, niobium, Ni–Cr alloy, palladium, platinum, silicon, tantalum, titanium or vanadium. The thinner layers are of the same types that are used for transient liquid phase bonding and are typically between 1 and 10μm thick; or between 1 and 20% of the thickness of the refractory core.

The main differences between transient liquid phase and partial transient liquid phase bonding are that the multi-layers used in partial transient liquid phase bonding, when in the liquid state, have to diffuse into the refractory core rather than into the materials to be joined in order to induce isothermal solidification. The liquid phases must also wet the ceramic substrates in order to ensure a strong bond, but this can be difficult due to the chemical inertness of ceramics. Some of the liquid that is formed from the thin layers will react with the ceramic substrate and add to the critical thickness.

As the temperature of the bond is raised to the melting point of each thin layer, both layers diffuse into the refractory core. Despite the small amount of liquid that forms initially, the liquid melts the refractory core back quite dramatically during additional heating, due to a concave liquidus. The melt-back continues until the assembly has attained bonding temperature. Meanwhile the liquid formed widens so as to approach the width of the original thin layer; due here to a convex liquidus. Isothermal solidification then occurs on both sides of the multi-layered interlayer. Isothermal solidification can in fact be complete for one side while the other liquid region has only partially solidified.

It will be recalled that the isothermal solidification and homogenization times for transient liquid phase bonding depend upon elements diffusing into the effectively-infinite materials which are to be joined. In the case of partial transient liquid phase bonding, the elements tend to have lower diffusivities. However, the maximum diffusion path is of the order of 100μm; thus leading to similar bonding times. The so-called self-contained nature of partial transient liquid phase bonding multi-layers means that the latter combine the benefits of brazing and diffusion bonding. The lower bonding temperatures can limit thermally-induced stresses and unwanted intermetallic reactions. Diffusion occurs here at a smaller scale (circa 100μm), so that bonding using slowly diffusing elements can take place within an acceptable time frame.

Materials Research Forum LLC
doi: http://dx.doi.org/10.21741/ 9781644900055

Applications of the Process

It is only natural to perceive a sort of 'hierarchy' in the joining of materials. That is, joining two objects having exactly the same nature tends to be seen as the easiest task, even though the welding of certain nickel alloys offers an immediate counter-example. Nevertheless, one cannot avoid imagining a progression of difficulty in going from the joining of a metal to itself, to joining an alloy to a similar alloy, thence perhaps to bonding apparently immiscible metals or alloys … and finally to bonding materials from entirely different groups, as when joining a metal to a ceramic. It is interesting to see how the present method can handle all of these cases.

Self-Bonding of Pure Materials

Metals

Aluminium

Pure aluminium was bonded by using an aluminium-based interlayer and heating to 595C for 120s in flowing argon under a pressure of 7MPa. An homogeneous bonding zone was observed in the joint, and the latter contained silicon oxides and voids as defects. The tensile strength of the joint was 190MPa, and no failure occurred when the joint was bent through 180°, indicating a strength and ductility equal to those of the base metal[13]. Dissolution and isothermal solidification during the transient liquid phase bonding of aluminium using a pure copper interlayer were modelled numerically on the basis of a diffusion-controlled process. Changes in volume due to interdiffusion between the aluminium and copper, and to the solid-liquid transformation were accounted for. The effect of an applied load was examined by setting up a simple force balance between the surface and interface energies of the base metal and the liquid which formed in the bonding region[14]. The predicted early stages of dissolution agreed with experiment, as did the predicted isothermal solidification time under an applied load.

Copper

Copper could be bonded below 250C under a pressure of 0.1MPa by using an interlayer of mixed copper nanoparticles and Sn-Bi eutectic powder[15]. A numerical simulation of the solute-induced melting of added powder particles during transient liquid-phase bonding, taking account of rapid interfacial processes occurring simultaneously at several moving liquid|solid interfaces, has shown that solute transport into the added powder particles during equilibration of the liquid composition is an important factor which affects the particles' melting behavior. The solute-transport dependence of the degree of particle melting affects the kinetics of solid-state solute diffusion within the particle. These results aid the design of interlayer powder mixtures containing base-alloy particles[16]. These formulations can produce monocrystalline joints having crystallographic orientations which match those of monocrystalline substrate materials. They also permitted the use of much shorter processing times than previously thought possible. Decreasing the temperature for copper mixed with 65wt% of Sn-Bi led to a decreased shear strength. For a treatment temperature of 200C, the highest shear strength was greater than 20MPa. It was essential to use copper nanoparticles in order to accelerate reactions so that none of the initial Sn-Bi remained after processing. Liquid formed at about 196C by reaction between newly-formed Cu_6Sn_5 and the bismuth phase was expected to aid joint densification and strengthening. This liquid could solidify as hypereutectic Sn–Bi but, upon heating at 200C, re-melting at 139C was not observed[17]. It was assumed that the proportion of solidified Sn–Bi eutectic in copper mixed with 65wt% of Sn-Bi at 200C was so small that, when reheated at 150C, the shear strength was equivalent to that at room temperature. In a similar manner, copper was bonded by using an interlayer of Sn–Bi eutectic plus silver particles and heating at 250C under a pressure of 0.02MPa in a reducing environment. The shear strength of joints made using 30wt% of added Sn–Bi was greater than 20MPa. The formation of intermetallic compounds was thought to strengthen the interface and matrix[18]. The re-melting temperature changed, from the Sn–Bi eutectic temperature, to about 262C. It was also bonded by using a tin interlayer and applying ultrasonics. In the usual non-ultrasonic process, Cu_3Sn always nucleated at Cu_6Sn_5|Cu interfaces and grew - at the expense of the Cu_6Sn_5 phase - towards the joint centre. This eventually led to a completely intermetallic joint which consisted of a single Cu_3Sn phase having columnar grains. When using ultrasonics, Cu_3Sn nucleated and grew randomly within the entire joint, leading to an intermetallic joint which consisted of a single Cu_3Sn phase having equiaxed grains[19]. This difference was attributed entirely to the effect of ultrasonics upon the nucleation and growth mechanisms of Cu_3Sn. The resultant joints could comprise markedly refined Cu_6Sn_5 grains, with an average size of 3.5μm. The joints also tended to exhibit more uniform mechanical properties, with elastic modulus and hardness values of about 123GPa and 6.0GPa, respectively, and a shear strength of 60.1MPa[20]. Copper has also been bonded by using a Sn-Ag powder interlayer and heating at 260C for only 600s. The liquid tin in the interlayer was completely consumed, and the joint shear strength could attain 39.5MPa. The Ag/Sn ratio was an important factor. Bonding using Sn70Ag yielded the highest shear strength of 72.3MPa, and gave a dense microstructure with few voids. The size of SnAgCu powders affected the void size in the joint, and the size of silver powders markedly affected the reaction process, suggesting that powders which are too large are not desirable[21]. The oxidation of small tin powder was a possible problem. The use of mixed powders was better than that of foil in that the bonding time was much shorter and the shear strength much higher, due to the void size and distribution. It was similarly bonded by using a Au–Sn interlayer. During bonding, $(Au,Cu)_5Sn$ formed first, followed by the appearance of new $\alpha(Au)$ and $Au_{6.6}Cu_{9.6}Sn_{3.8}$ phases and the final formation of a combination of $\alpha'(Au)$, $\alpha(Au)$ and $Au_{6.6}Cu_{9.6}Sn_{3.8}$ phases when the Au–Sn interlayer was exhausted. Volume contraction due to the consumption of the interlayer led to pore formation and to consequent shear-strength deterioration of the Cu|Au-Sn|Cu joint.

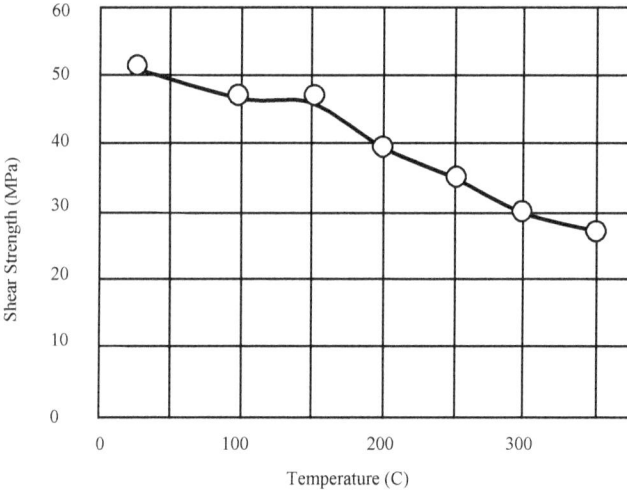

Figure 8. Shear strength of bonded Cu|Au–Sn|Cu joints as a function of the test temperature

The presence of the $Au_{6.6}Cu_{9.6}Sn_{3.8}$ also tended to reduce the shear strength. The overall shear strength of a joint without pores was about 50MPa. The shear strength was still 28MPa, even at 350C (figure 8), although this temperature was 70C higher than the melting-point of the interlayer alloy[22]. The shear strength remained at 50MPa, even after heating to 250 or 350C for 400h. It decreased slightly only after 400 cycles (figure 9). When a thin interlayer of pure tin foil was sandwiched between two pieces of copper foil and heated to 260, 300 or 340C in forming gas for 300s to 8h, the interfacial microstructures exhibited Cu_6Sn_5 growth and Cu_3Sn columnar crystals. There was a marked difference in the thicknesses of the Cu_6Sn_5 and Cu_3Sn layers which formed at the two original boundary planes in the Cu|Sn|Cu samples[23]. There was also a form of grain-boundary/molten-channel controlled growth of Cu_6Sn_5, with a time dependence that was similar to that for volume-diffusion controlled growth. When pure copper was bonded by using a Cu-46Zr-8Al amorphous alloy foil as an interlayer and assemblies were heated to 750 to 900C under 0.5MPa for 300s, the joint interface was satisfactory and the tensile strength after bonding at 900C was 345MPa, with good plasticity.. When the bonding temperature was 750 to 800C, diffusion reaction occurred at the interface and led to the formation of three Cu-Zr intermetallic layers[24]. When the bonding temperature was 850 to 900C, the copper matrix grew into the interlayer in cellular dendritic form and a lamellar eutectic structure (Cu_9Zr_2|copper) appeared in the interlayer.

A study was made of the critical interlayer thickness which is required in order to avoid pore formation during bonding. Pores are a result of the growth and subsequent impingement of Cu_6Sn_5 intermetallic grains from the two bonding surfaces before formation of the transient liquid phase. A criterion for the critical interlayer thickness could be based upon the size of the largest intermetallic grain[25]. A high heating-rate had a beneficial effect upon both the critical interlayer thickness and upon the bonding time required to produce a desired microstructure.

The strength and toughness of joints produced by bonding in the Cu-Sn system was studied by using a substrate consisting of oxide-dispersion strengthened copper. Three microstructures were considered: a uniform layer of δ-phase, $Cu_{41}Sn_{11}$, a two-phase microstructure comprising the δ-phase and a dispersion of ductile copper particles, and a uniform copper solid solution. The δ-phase exhibited a moderately high strength of 300MPa, but a toughness of only about 5MPa√m. The addition of copper particles increased the strength and toughness by about 30%, and this could be explained in terms of familiar models of ductile-phase toughening. Conversion of the intermetallic into copper solid solution led to a decrease in strength to 200MPa, but an increase in toughness to 13MPa√m. The latter effect was attributed to a reduction in the flow stress of the joint material[26]. The conversion to copper was also accompanied by the formation of voids; mainly near to the previous boundary between the δ-phase and adjoining copper. Such voids were expected to impair the joint properties.

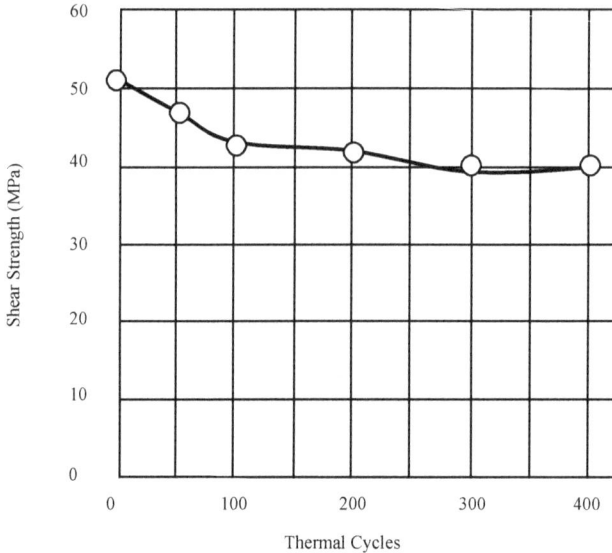

*Figure 9. Shear strength of bonded Cu|Au–Sn|Cu
joints as a function of thermal cycling*

The interactions which occur in a Cu|Sn|Cu structure during transient liquid phase bonding conditions have been studied by using the multiphase-field method, especially with regard to the appearance of η-Cu_6Sn_5 and Cu_3Sn. It was found that, while molten tin was present, growth of η-phase layers predominated. When the molten tin had disappeared, the η-phases of the upper and lower layers impinged and large grains continued to coarsen at the expense of small grains. Growth of the layers simultaneously accelerated and continued until the η-layer was completely used up or the reaction ceased. Such simulations were in good qualitative agreement with the morphologies of intermetallic grains[27]. There was good quantitative agreement with regard with regard to changes in the intermetallic layers.

Gold

Gold can be bonded by using an indium interlayer. The sequential interfacial reactions between gold and indium were investigated at 190C for up to 0.5h under pressures of 0.2 to 10MPa. The joint was fully converted into gold-rich Au_7In_3 and, during bonding, the indium interlayer was sequentially transformed in the order: Au_7In_3 + AuIn + $AuIn_2$, then Au_7In_3 + AuIn and finally the Au_7In_3. Good joints were formed by using pressures of 5 and 10MPa for short times[28].

Iron

Commercially pure iron was bonded by using 25μm copper or 100μm Au-12Ge eutectic interlayers and heating at 1000 to 1090C for 10 to 30h. The copper interlayer did not diffuse completely into the base metal, and the compositions of the residual interlayer and base metal at the interface were limited by the solubility ranges of copper and γ-iron. A residual copper interlayer was also observed after bonding at 1100C for 3 to 4h, but no residual copper was found after 5h at 1100C. On the other hand; a porous microstructure was observed along the joint center-line. The copper diffused mainly along austenite grain boundaries under all of the bonding conditions. The iron was also bonded by using a Au-12Ge interlayer at 900 to 990C for 1 to 15h. After 1h at 900C, the residual interlayer thickness was 12.6μm. This decreased to 6μm when the time was increased to 10h, and a fingerprint-like microstructure was observed in the residual interlayer. After 15h at 920C, the interlayer had completely diffused into the base metal in some areas but was retained in some areas[29]. The thickness of the residual interlayer was then 1.25 to 3.8μm, and gold-rich particles were dispersed in the base metal near to the interface. The highest ultimate tensile strength obtained by bonding using a copper interlayer and heating at 1030C for 10h was 291MPa. The highest strength obtained by using a Au-12Ge interlayer and heating at 950C for 15h was 315MPa. An attempt was made to use a liquid, with solid particles at the bonding temperature, as an interlayer. The growth of solid particles in Fe-4.5wt%P and Fe-1.16wt%B interlayers when bonding pure iron was linearly dependent upon the square-root of time. The slope of this dependence was higher than that expected for the usual bonding method and this was attributed to the contribution made by the solid particles distributed throughout the interlayer. Those solid particles exhibited no growth[30]. When pure iron particles coexisted with melt of equilibrium composition, such particles grew very rapidly.

Magnesium

The bonding of magnesium by using a zinc interlayer and heating at 370C in air was facilitated by using ultrasonics. The latter tended to accelerate eutectic reaction between the magnesium and zinc as well as accelerating isothermal solidification. The joint shear strength increased non-linearly with the ultrasonic exposure time, and this was related to the thicknesses of $Mg_{51}Zn_{20}$ and MgZn layers, the α-Mg(Zn) solid-solution layer and the overall joint extent. The highest average shear strength of joints made under 120s of ultrasonic exposure attained 106.4MPa; almost equal to the strength of the magnesium base metal. In other tests the shear strength reached 109.3.MPa; exactly equal to that of the base metal. As well as the previously mentioned effects, it was noted that surface oxide films were removed by the ultrasonics[31]. The intermetallic compounds in the joint decreased with increasing bonding temperature, and a fully solid-solution joint without intermetallics or pores was obtained by using a two-step process: ultrasonics at 370C and then at 490C. The time required for isothermal solidification was thereby shortened to a few seconds due to the squeezing-out of liquid, and accelerated diffusion[32].

Nickel

Specimens of the form, Ni|B|Ni, were bonded at 1433 to 1473K under vacuum. Liquid metal naturally formed in the bonding zone, due to intermixing of atoms diffusing from the solid interlayer and the base metal, following an incubation time of 25s. The initial width and maximum width of the liquid layer were 3.5 and 9 times, respectively, wider than the thickness of the intermediate layer[33]. It was mentioned that pure carbon could be used as an interlayer, but that choice does not seem to have been widely investigated. In the joining of nickel using ternary Ni-Si-B interlayers, it was suggested that the diffusion of boron out of the liquid and into the solid substrate, before equilibration of the liquid and solid phases, could result in the development of appreciable boron concentrations in the substrate. This, in turn, led to the precipitation of boride phases in the substrate during holding at temperatures which were below the binary nickel-boron eutectic temperature. The formation of boride phases during holding at the bonding temperature was not predicted the usual models for the process, and the borides were not removed by prolonged holding at the bonding temperature[34]. They could thus affect the properties of the joint. Bonding above the binary nickel-boron eutectic temperature caused localized melting of the substrate. This liquid re-solidified during prolonged heating and did not result in the formation of persistent boride phases. When bonding using an interlayer containing phosphorus as a melting-point depressant, an insufficient holding time for complete isothermal solidification of a molten interlayer resulted in the formation of eutectic-type microconstituents along the joint centreline region. An increase in the eutectic thickness with increasing interlayer gap-size was observed in joints prepared at a given temperature and holding time. The eutectic acted as a preferred continuous path for crack initiation and/or propagation. In some cases, an increase in bonding temperature resulted in the formation of thicker centreline eutectic[35]. In other cases, avoidance of the eutectic was achieved in a relatively shorter time at a lower temperature. The effect of the base-metal grain-size on isothermal solidification using a Ni-11wt%P interlayer has been investigated. Monocrystalline, coarse-grained and fine-grained nickel was heated at

1423K for various holding times and air-cooled, oil-quenched or water-quenched. The eutectic width in air-cooled and oil-quenched samples was smaller than that in water-quenched ones. This was attributed to differing solidification modes during cooling. The eutectic width decreased linearly with the square-root of the heating-time in all cases. The completion time for isothermal solidification decreased in the order: monocrystalline, coarse-grained, fine-grained[36]. The order was attributed to effect of grain boundaries upon the diffusion coefficient of phosphorus in solid nickel[37]. Bonding using a Ni-11wt%P interlayer has been simulated by using a phase-field model and a moving boundary model. Dissolution of the base metal and isothermal solidification were simulated. The results for isothermal solidification were the same using either model. The change in concentration at the solid|liquid interface during dissolution of the base metal was predicted to exhibit a deviation from the local equilibrium concentration when heating at a high rate, according to the phase-field simulation[38]. According to the other model, local equilibrium was always maintained. Another model for the bonding process was based upon a diffusional analysis. In this model, diffusion-controlled transformation was assumed to occur, and base-metal dissolution in the early stages and the subsequent isothermal solidification stage were simulated. The calculated width of the eutectic structure in the bonding region agreed well with experiment. The simulation showed that the isothermal solidification time markedly decreased with increasing partition coefficient of the solute element[39]. This was confirmed by using copper as the interlayer, given that the partition coefficient of copper in nickel is close to unity. The copper interlayer did indeed sharply shorten the isothermal solidification time. As a further exploration of the effect of grain size, a two-dimensional finite-difference model was used to analyze the effect of grain boundary regions upon the migration of the liquid|solid interface when using a Ni-11wt%P interlayer. The model took account of the case where the grain boundary intersected the liquid|solid interface at right angles, and considered solute diffusion in the solid and liquid, a higher diffusivity in the grain boundary region, the balance between grain-boundary energy and liquid|solid interfacial energy, the interfacial energy due to the curvature of the liquid|solid interface and diffusional flow along the liquid|solid interface which was produced by a gradient in solute chemical potential. The increased solute diffusivity in the grain boundary region had a negligible effect upon the migration of the liquid|solid interface in the bulk region, and was predicted to shift the interface in the grain boundary region in a direction which was opposite to that observed experimentally[40]. On the other hand, when other factors were considered the liquid|solid interface in the region of the grain boundary was predicted to be displaced in the direction observed experimentally and liquid penetration which was comparable to that observed occurred in the grain boundary region. Experiments on cast nickel with a grain size of >4mm and on fine-grained nickel with a grain size of about 40μm showed that the rate of isothermal solidification was greater when fine-grained nickel was bonded at 1200C. Liquid penetration at the grain boundaries accelerated isothermal solidification by increasing the effective solid|liquid interfacial area and by increasing the rate of solute diffusion into the base material. Random high-angle boundaries also had a greater effect upon the rate of isothermal solidification than did ordered boundaries; including low-angle or twin boundaries[41]. In a further model for dissolution and isothermal solidification during bonding using a Ni-15.2Cr-4.0wt%B interlayer, the interface velocity was deduced from the mass balance of solutes diffusing into and out of the interface. When nickel with a 30μm-thick interlayer was kept at 1473K for 1h, isothermal solidification was almost completed, but when it was kept at 1373K for 1h, residual liquid remained in the bonding region[42]. The solidification sequence of the residual liquid during cooling was deduced from the Scheil simulation. This predicted that solidification of the residual liquid finished with a ternary eutectic reaction, $L \rightarrow fcc + Ni_3B + CrB$. The calculated width of the eutectic region agreed with experiment. Nickel could also be bonded by using a Sn-30Ni interlayer and heating at 340C for 4h. The bond comprised mainly Ni_3Sn_4 plus residual nickel particles. The bonding layer was stable, and its strength did not change noticeably with aging time at 350C because of the difficulty of nucleation of Ni_3Sn and the relatively slow growth of Ni_3Sn_2. When the aging temperature was increased to 500C, the residual nickel particles gradually dissolved and the bonding layer formed a stable structure comprising predominantly Ni_3Sn_2 after 36h. Voids formed due to a 4.43% volume shrinkage resulting from Ni_3Sn_2 formation[43]. The shear strength first increased due to the Ni_3Sn_2 formation, but then decreased slightly due to the voids. Following aging (500C, 100h), the shear strength was still 29.6MPa. During heating at 300C for 4h and at 340C for 3h, the tin completely transformed into Ni_3Sn_4. When the Ni_3Sn_4 around nickel particles was pressed together, the porosity of the bonding layer anomalously increased. A volume shrinkage of 14.94% at 340C occurred when nickel reacted with tin to form Ni_3Sn_4; leading to void formation[44]. The hardness of the bonding layer after 4h at 340C was uniform, with an average value of up to 3.66GPa; close to that of bulk Ni_3Sn_4. Shear tests showed that, as the holding time was increased from 1 to 3h at 340C, the difference in shear strength between room temperature and 350C was large because of the presence of tin[45]. When the holding time was 3h or longer, the tin had been completely transformed into Ni-Sn intermetallics and the strength was similar at room temperature and 350C.

Silver

In an early study, a model Ag|Cu|Ag sandwich joint and a simple eutectic phase diagram were used to study the various bonding stages. The results confirmed that transient liquid phase bonding is a diffusional process which occurs in distinctive stages. The two most important of these are the widening and homogenization of the previously dissolved liquid interlayer, and subsequent solidification and shrinkage of the interlayer. The first stage involves diffusional processes in the liquid phase and in the adjoining solid. The second is controlled mainly by diffusion in the solid phase. The model indicated that, in most eutectic systems, there is an optimal bonding temperature, corresponding to the shortest time required for complete solidification[46]. Study of a Ag|Ag-20wt%Cu|Ag sandwich joint confirmed that the use of an interlayer alloy which was close to the eutectic composition markedly shortened the process. The kinetics of the interfacial reaction of a thin layer of tin sandwiched between two pieces of silver foil were determined at 260, 300 and 340C, and a time-dependence of the form, $t^{1/3}$, was deduced for the kinetics of consumption of the remaining tin and for the thickening of the Ag$_3$Sn which formed between tin and silver[47]. The results were explained in terms of a model for the grain-boundary/molten-channel controlled growth of intermetallic compounds. In this case, silver diffusion through the molten channels which existed between the previously formed Ag$_3$Sn was the controlling kinetics mechanism. Silver was again bonded by using a tin interlayer, but with the addition of ultrasonics. Rapid consumption of the transient liquid phase was attributed to an accelerated dissolution of the silver substrate and to the extrusion of liquid tin by the ultrasonics[48]. The ultrasonics led to an elongated morphology of the Ag$_3$Sn grains which developed during Ag|Sn interfacial reaction, together with a widening of the grooves between neighboring grains which provided stable molten channels for silver atom diffusion from the substrate.

Tin

In the case of solid-state diffusion-bonding without an interlayer, copious oxide inclusions were distributed within a 100nm-wide range along the bond interface. On the other hand, a liquid phase which was created by eutectic reaction of a bismuth interlayer with the tin substrate decreased the width of the interfacial region containing oxide inclusions. A layer structure formed which was a few 100nm thick and consisted mainly of SnO$_2$. The melt also enhanced annihilation of the non-contacting areas of the interface. The oxide layer became discontinuous, and coalesced, with increasing bonding temperature and pressure[49]. The areas where no oxide inclusions were present at the interface increased when the molten phase formed and thus the bond strength increased at lower bonding temperatures and pressures when a bismuth interlayer was used.

Titanium

Commercially pure titanium was bonded by using a silver-based AMS4772 interlayer and heating at 900 to 1000C under a vacuum of 6 x 10^{-7}Torr. The tendency to form an isothermally solidified joint increased with increasing bonding time. No sign of athermal solidification was found when heating at 1000C for 1.5h. Interlayer elements such as silver and copper were more uniformly distributed in fully isothermally solidified bonds, while there was considerable aggregation of those elements in athermally solidified bonds. The highest shear strength was found after 1.5h at 1000C[50].

Ceramics

Alumina

Alumina was originally joined to alumina at 1150C by using Cu|Ni|Cu interlayers which formed a thin layer of copper-rich transient liquid phase. This led to bonding by a greater than 94at%Ni interlayer. The flexural strength of as-bonded samples ranged from 61 to 267MPa, with an average of 160MPa and a standard deviation of ±63MPa. The highest flexural strength was found when failure occurred in the ceramic. Post-bonding annealing (1000C, 10h), in air or argon, decreased the average room-temperature strength to 138 or 74MPa, respectively[51]. The formation of spinel was thought to contribute to the scatter of the results of as-processed samples and to the decrease in strength following annealing. Similar joining of alumina by using multilayer Cu|Pt filler was achieved at 1150C, giving a platinum-rich interlayer[52]. The ceramic|metal interface strength exceeded that of the ceramic. Post-bonding annealing (1000C, 10h), in air or argon, had differing effects upon the room-temperature joint strength. The effect of chromium, nickel and Ni-20Cr alloy additions upon the wetting characteristics of liquid copper on Al_2O_3 was determined by using the sessile drop method in a vacuum at 1150C. This led to the bonding of Al_2O_3 in vacuum at 1150C by using micro-designed multilayer Cu|Ni-20Cr|Cu interlayers[53]. This demonstrated that the solution of chromium in a liquid copper film improved the wettability of liquid copper film on Al_2O_3 and aided the formation of strong Al_2O_3 bonds upon using a Cu|Ni-20Cr|Cu interlayer. Alumina pieces coated with boron oxide layers of various thickness have been bonded by heating in air at 800C for various times under modest pressure. The pure boron oxide melted at low temperatures and reacted with alumina to form a stable high melting-point compound. As usual, this transient liquid phase bonding had the advantage of producing a highly temperature resistant ceramic joint by using low processing temperatures. The maximum flexural strength attained 155MPa after heating at 800C for 15h, using a 21µm-thick interlayer[54]. When the alumina was coated with 3µm of boron oxide, the maximum bending strength was

Materials Research Forum LLC
doi: http://dx.doi.org/10.21741/ 9781644900055

71MPa after heating at 800C for 0.25h. The presence of $3Al_2O_3$-B_3O_3, $2Al_2O_3$-B_3O_3 and $9Al_2O_3$-$2B_3O_3$ was noted in the joint, and $2Al_2O_3$-B_3O_3 whiskers predominated in the joint which exhibited the highest strength[55]. Alumina was joined by using a multilayer In|Cu-Ag|In filler and heating at 500, 600 or 700C. Use of the thin indium layers allowed the system to bond at much lower temperatures than those normally used for brazing with Cu-Ag eutectic[56]. Bonds made at 500, 600 or 700C, using holding times ranging from 1.5 to 24h, exhibited fracture strengths that were greater than 220MPa. Analysis of the interlayers showed that neither Ag-In nor Cu-In intermetallic phases were formed. The time required to bond high-purity Al_2O_3 could be markedly reduced by using Ni|Nb|Ni multilayer fillers which formed thin transient-liquid films[57]. Strengths in excess of 500MPa could be obtained while using isothermal holding times as short as 300s at 1400C. The multilayer structure provided liquid, at reduced temperatures, which then filled the interfacial gaps and accelerated the disappearance of liquid[58]. Interlayers having re-melt temperatures greater than 2200C could be obtained within 300s at 1400C. During room-temperature bend tests fracture began, and continued, almost entirely within the ceramic[59]. High-purity Al_2O_3 was also joined by using multilayer metallic interlayers which were niobium- or vanadium-based; both being designed to produce a thin transient liquid-phase layer[60]. The Ni|Mo|Nb|Mo|Ni interlayers reduced residual stresses and improved the ceramic/metal interface microstructure; thus creating better joints. Alumina has more recently been joined to alumina by using a thin layer of bismuth oxide as a interlayer and heating at 900, 1000 or 1100C for various times. Longer times resulted in better mechanical properties, and the highest joint strength of about 80MPa was obtained by heating at 900C for 10h[61]. Alumina has also been bonded by using a mixture of aluminium powder and silica powder as an interlayer. Chemical reaction of aluminium with the silica occurred in the interlayer to produce alumina and silicon. The material could also be subjected to cold isostatic pressing before bonding in order to improve the bond strength. Bonding occurred when the interlayer Al/SiO_2 ratio was 1:0.84 or 1:0.42, but did not occur for a ratio of 1:1.67. When the material was cold isostatically pressed before bonding, the bonds were far stronger. This was true for temperatures ranging from ambient to above the melting point of aluminium. In bend tests, fracture occurred at the boundary between the alumina matrix and the interlayer at room temperature[62]. It occurred in the interlayer at temperatures above the melting point of aluminium.

Carbon

Carbon/carbon composites can be bonded by using a Ti-Ni-Al_2O_3-Si compound as an interlayer. A graded interface and a core interlayer form in the carbon|carbon joints via diffusion, chemical reaction and hot-press sintering. A liquid phase infiltrated the C/C matrix and exerted a so-called nail-effect, which improved the mechanical properties. The average joint shear strength attained 21.30MPa at room temperature, and this was maintained at up to 1273K[63]. The shear strength slowly decreased during thermal cycling from room temperature to 1373K. Following 30 such cycles the retention of strength still amounted to 73.10%. Two-dimensional fibre-reinforced carbon/carbon composites have also been bonded by using a Ti|Ni|Ti sandwich interlayer and heating at 1050C for 20 to 100min under an axial pressure of 0.1MPa. As before, a Ti-Ni molten eutectic infiltrated the interconnected capillaries of the carbon/carbon composite via the open pores of the composite bonding surface. The existence of breaks in the alloy, and reaction between the titanium and carbon were surprisingly beneficial in improving the bond strength[64]. The shear strength of the joint could be as high as 37.4MPa.

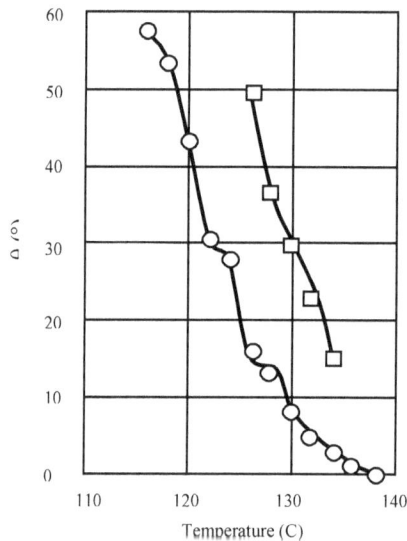

Figure 10. Apparent contact angle of molten Ni-Nb alloy on HfB₂-MoSi₂ composite as a function of temperature Circles: pure nickel, squares: Ni-40at%Nb

Hafnium Boride

The wetting behavior of molten Ni-Nb alloy that formed during bonding, when a Ni|Nb|Ni interlayer was used to bond HfB_2 with $MoSi_2$ additions, was systematically investigated at 1500C. Pure nickel melted at a temperature which was far below its

Materials Research Foundations 43 (2019)   doi: http://dx.doi.org/10.21741/ 9781644900055

normal melting-point during heating and this was largely attributed to interfacial reaction between HfB_2, $MoSi_2$ and the droplets.

The contact angles of all of the Ni-Nb alloys decreased rapidly with time and exhibited good wetting (figure 10) of the HfB_2. Pure nickel droplets disappeared entirely at about 1400C during heating. The reaction layer contained several intermetallics involving silicon; suggesting that the sintering aid, $MoSi_2$, was implicated[65]. With the addition of 40at%Nb, there was less interfacial reaction at the Ni-Nb|HfB_2 interface.

Hafnium Carbide

The bonding of undoped and $MoSi_2$-doped HfC can be performed by using a Ni-Nb-Ni multilayer filler and heating at 1400C in a high-vacuum furnace for 0.5h. The reaction layer at the interlayer|ceramic interface contained mixed carbides and, depending upon the ceramic, Ni-Nb-Hf, Ni-Nb-Hf-Si or Ni-Nb-Si alloys. Nano-indentation traverses of the reaction layer between the bulk ceramic and the niobium foil mid-plane revealed a clear transition zone, across which the indentation modulus and hardness varied. Crack-free joints have been obtained by using undoped HfC, while the addition of 5vol%$MoSi_2$ introduced isolated cracks, less than 5μm long, within the reaction layer[66]. The addition of 15vol%$MoSi_2$ led to widespread cracking throughout the reaction layer. If the latter exceeded a critical thickness, as when 15vol%$MoSi_2$ was added, the residual stresses were high enough to cause extensive cracking and joint failure.

Molybdenum Disilicide

This material can be bonded by using a Cusil|Zr|Cusil interlayer and heating at 830 to 930C for 1h in an inert-gas tube furnace. Cusil is a commercial Cu-Ag eutectic composition. Heating at 830C produced a uniformly dense joint which consisted of various reaction phases and exhibited excellent bonding within the interfaces. Diffusion of silicon into the liquid phase transformed the $MoSi_2$ into Mo_5Si_3 containing some silver in solid solution, while pure silver and $Cu_{10}Zr_7$ were present in the interlayer[67]. Use of a higher temperature did not produce an adequate strength.

Silicon Nitride

When this nitride was bonded by using Au|Ni-Cr|Au multilayers, the Ni-Cr foil was attached by a thin CrN reaction-product layer while the gold dissolved to form a nickel-rich solid solution[68]. Joints having a room-temperature flexural strength of 272MPa were produced by heating at 1000C for 4h. Use of Ti|Ni|Ti interlayers led to interfacial structures of the form, Si_3N_4|TiN|Ti_5Si_3 + Ti_5Si_4 + Ni_3Si|(NiTi)|Ni_3Ti|Ni, after bonding. The activation energies for TiN layers and for mixed reaction layers, Ti_5Si_3 + Ti_5Si_4 +

Ni_3Si, formation were 546.8 and 543.9kJ/mol, respectively[69]. During the bonding, not only was there growth of the reaction layers, but also changes in the diffusion path due to marked variations in the concentration on the metal side. When the bonding was performed at 1000C using Ti|Cu|Ti multi-interlayers, Cu-Ti transient liquid alloy formed on the nitride surface and reacted to form a graded Si_3N_4|TiN|Ti_5Si_3 + Ti_5Si_4 + $TiSi_2$|$TiSi_2$ + Cu_3Ti_2(Si)|Cu interface. Due to variations of concentration in the transient liquid, the diffusion path during subsequent reaction changed[70]. The joint strength was affected by the thickness of the reaction layer and by the level of residual thermal stresses; which were in turn affected by the bonding time. When the nitride was instead bonded using Ti|Cu|Ni interlayers, the effect of the titanium foil thickness upon the bond strength was related to the thickness of the reaction layer. The given bonding parameters had no effect upon the thickness of the reaction layer at the Si_3N_4|Ti|Cu|Ni interface, but affected the bond strength by altering the residual stresses in the joint[71]. The microstructure of the Si_3N_4|Ti|Cu|Ni interface was: Si_3N_4|reaction-layer|Cu-Ni solid solution with a small amount of Cu-Ni-Ti|Ni. The nitride could also be bonded by using Ti|Cu|Ti multilayers, as above[72]. When interlayers of the form, Ti(5µm)|Cu(70µm)|Ti, were used the growth of the reaction layer and the movement of the isothermal solidification interface obeyed a parabolic law and was controlled by diffusion of the elements[73]. The joint strength increased with increasing holding time but, when this was longer than 25min, further increases led to a decrease in the joint strength. The latter was controlled by the thickness of the reaction layer[74]. The nitride could also be joined by using just a Ti|Ni interlayer, and heating at 1273 to 1423K under vacuum. The interfacial structure now consisted of Si_3N_4|TiN|Ti_5Si_3 + Ti_5Si_4 + Ni_3Si|(NiTi)|Ni_3Ti|Ni, and the NiTi layer was gradually consumed, with simultaneous growth of the reaction layer and the Ni_3Ti layer. The room-temperature joint strength was greatly affected by the reaction-layer thickness, while the high-temperature joint strength depended markedly upon whether a low melting-point NiTi layer was present in the joint[75]. Strengths greater than 100MPa could be maintained at up to 800C when the NiTi layer was completely consumed. In general, multilayers of the form, Ti|Cu|X, could be used where X was nickel, platinum, gold, palladium, etc., but X and copper had to be mutually soluble in the solid and in the liquid[76]. Double partial transient liquid phase bonding of Si_3N_4 ceramic was achieved by using Ti/Cu/Ni interlayers. The room-temperature bond strength was increased by using a second bonding temperature and a second holding time. The bonding parameters had a slight effect upon the thickness of the reaction layer in the Si_3N_4|Ti|Cu|Ni interfaces of double partial transient liquid phase bonds. The bond strength was greatest at 400C, and decreased with increasing temperature, but the high-temperature strength exhibited good stability when the temperature was less than 800C[77].

Ti$_3$SiC$_2$

The bonding of layered ternary Ti$_3$SiC$_2$ ceramics was achieved by using an aluminium interlayer and heating at 1100 to 1500C for 2h under a 5MPa load in argon. It was found that Ti$_3$Si(Al)C$_2$ solid solutions rather than intermetallic compounds formed at the interface. The bonding was attributed to the diffusion of aluminium into the Ti$_3$SiC$_2$. The maximum flexural strength attained 263MPa, about 65% of that of Ti$_3$SiC$_2$, in samples heated at 1500C for 2h under 5MPa[78]. This strength was maintained at up to 1000C.

Zirconium Carbide

Good bonding of ZrC ceramics can be achieved by using a Ni|Nb|Ni multilayer filler at 1673K[79].

Bonding Different Pure Materials

Metals

Aluminium|Copper

Aluminium and copper were bonded without using any interlayer, or by using an Al-11Si-4Cu-2Mg or Al-4.5Si-2Cu-1Mg foil. The bonding was performed by using sandwich assemblies at various temperatures under vacuum. Two layers were formed at the interfaces of all of the joints; the main difference in the joints being the microstructures of the Al$_2$Cu and Al$_2$Cu|Al interfacial structures. The interface of the joint made without using an interlayer consisted mainly of Al$_2$Cu+Al eutectic near to the aluminium side, in which the Al$_2$Cu phase had a continuous structure. The formation of Al$_2$Cu could be greatly inhibited in the joints made by using an interlayer[80]. The shear strength of the joints made using Al-11Si-4Cu-2Mg foil at 575C could attain 77MPa; close to the shear strength of the aluminium base. The resistivity of the joint was close to the theoretical value. Induction diffusion bonding (600C, 2s, 9MPa) was also used to join copper and aluminium, using a foil interlayer. Failure during tensile testing occurred on the aluminium side. No failure occurred when the joint was bent through 180°. The interface contained Cu$_9$Al$_4$ and CuAl$_2$, had a total thickness of 2μm, and contained no voids or oxide scale in the join[81]. Step-wise vacuum diffusion bonding was used to join copper and aluminium using a nickel interlayer. The latter layer effectively prevented the creation of brittle intermetallic compounds in the copper and aluminium. Two-layer Al$_3$Ni and Al$_3$Ni$_2$ was observed in the nickel|aluminium interface, together with a solid solution having a composition gradient in the copper|nickel interface. When the nickel foil thickness was 20μm, the foil was entirely consumed, so obvious defects were absent from the interface and the interface shear strength attained its maximum value[82]. Aluminium/copper bimetals have been prepared by diffusion bonding at 450 to 550C after again depositing a nickel-based coating onto the pure aluminium, here via immersion plating. The nickel interlayer effectively eliminated the formation of Al-Cu intermetallic compounds: the aluminium/nickel interface consisted of Al_3Ni and Al_3Ni_2, while nickel-copper solid solution was found at the nickel|copper interface. The nickel interlayer improved the tensile shear strength of the joint, with a joint annealed at 500C exhibiting a maximum tensile shear strength of 34.7MPa[83].

Aluminium|Magnesium

When magnesium and aluminium were joined using vacuum diffusion bonding, the multi-layer diffusion couple at the join consisted of aluminium-based solid solution, an Al_3Mg_2 layer, an $Al_{12}Mg_{17}$ layer and a magnesium-based solid solution. The microhardness of the transition layer was higher than those of the aluminium or magnesium base metals, due to the presence of the intermetallic phases[84]. The shear fracture was brittle fracture and occurred at the joint interface. When the diffusion bonding of magnesium and aluminium alloys was performed at between 390 and 490C using a 30μm-thick silver foil interlayer, the joint structure was: Mg | Mg-Ag solid solution | Mg_3Ag | MgAg | Ag | Ag-Al solid solution | Ag_2Al | Al. The silver diffusion barrier prevented the formation of brittle intermetallic compounds between the magnesium and aluminium. Those intermetallics found at the joint interface included the more ductile ε-Mg_3Ag, β'-MgAg and δ-Ag_2Al. The shear strength of the join increased, with increasing bonding temperature, to a maximum of 11.8MPa at 470C and fracture occurred mainly in the Ag_2Al layer[85]. The ultrasonic-assisted transient liquid phase bonding of aluminium to magnesium was performed by using pure tin interlayers. The formation of intermetallics such as Al_3Mg_2 and $Al_{12}Mg_{17}$ was avoided, although Mg_2Sn formed in the joints. An optimum joint shear strength of 60MPa was obtained at 220C after 4s by using a suitable ultrasonic power[86]. The thickness of the Mg_2Sn was best decreased to less than 15% of the joint width. Failure of the joints occurred within the Mg_2Sn layer.

Copper|Nickel

Nickel and copper were bonded by using a tin interlayer and heating at 260, 300 or 340C. The grain morphology of an $(Cu,Ni)_6Sn_5$ intermetallic was closely related to the nickel concentration gradient across the joint and was markedly affected by the bonding temperature[87]. The respective shear strengths for the three bonding temperatures were 49.8, 50.3 and 42.7MPa. Micro-joints between copper and nickel were prepared by using

Cu|Sn(1.5μm)|Ni sandwich layers and heating at 240C and 290C. Two layers, a thick $(Cu,Ni)_6Sn_5$ layer below an upper thin $(Cu,Ni)_3Sn$ layer, made up a micro-joint formed at 240C after 0.25h. The intermetallic interlayer in micro-joints after 5 to 25min at 290C was much more uniform and homogeneous[88]. When nickel and copper were bonded by using a tin interlayer, homogeneous $(Cu,Ni)_6Sn_5$ joints were formed rapidly due to ultrasound-induced liquid. When using traditional methods, the intermetallic joints which formed consisted mainly of $(Cu,Ni)_6Sn_5$ plus some Cu_3Sn. The morphology of the $(Cu,Ni)_6Sn_5$ ranged from fine-and-rounded, through needle-like, to coarse-and- rounded in going from the nickel side to the copper side; in precise accord with the nickel concentration gradient across the joint[89]. On the other hand, using ultrasound-induced joining, the joints contained only $(Cu,Ni)_6Sn_5$ with a uniform rounded shape, and the nickel concentration gradient was markedly narrowed. Thermodynamic analysis indicated that the critical nickel concentration which determined the change from needle-like shape to coarse and rounded was about 11at% at a bonding temperature of 260C. Ultrasound-induced joints had a higher shear strength than did the traditional intermetallic joints; 61.6MPa as opposed to 49.8MPa. Shear fracture tended to occur in the region of $(Cu,Ni)_6Sn_5$ grains having a coarse rounded shape[90].

Copper|Silver

Copper and silver were bonded by using a Sn-Ag mixed powder interlayer. The microstructure consisted of various zones. With increasing silver content, the number of Ag_3Sn grains in an *in situ* reaction zone increased but their size decreased. The thickness of the intermetallic layers at both of the interfacial-diffusion reaction zones decreased. When the silver content was greater than 70%, a large number of silver particles remained and a few pores formed in the bonded layer[91]. The shear strength and microhardness of the joint first increased but then eventually decreased with increasing silver content, attaining maxima of about 35MPa and 70HVN, respectively. Differential scanning calorimetry was used to determine the interface kinetics in a solid/liquid diffusion couple in order to clarify the isothermal solidification stage which occurred during transient liquid-phase bonding of silver and copper by using a Ag-Cu interlayer. The results were in close agreement with a model which assumed boundary movement, but the accuracy of the predictions depended sensitively upon the estimated solute diffusivity[92]. It was concluded that simple analytical models can be used to predict accurately the kinetics of isothermal solidification in simple binary systems when the effects of grain boundaries could be neglected. Copper and silver were bonded by using a tin interlayer and heating at 260 to 340C. The copper and silver reacted independently with the molten tin, but the growth of intermetallics from one side was blocked by those coming from other side when Cu_6Sn_5 contacted Ag_3Sn. No ternary phase formed. Pores were distributed at the $Cu_6Sn_5|Ag_3Sn$ interface and between the grain boundaries when the residual tin had been fully used up. The pores gradually disappeared with continuing reaction. The shear strength of the joints increased with increasing bonding time, and the adhesion of the $Cu_6Sn_5|Ag_3Sn$ interface was weaker than that of the $Cu_3Sn|Ag_3Sn$ interface[93]. Cracks initiated in pores and propagated mainly along the $Cu_6Sn_5|Ag_3Sn$ interface as the joint consisted of layered Cu_3Sn, Cu_6Sn_5 and Ag_3Sn. The failure path passed only through the Cu_3Sn layer when the Cu_6Sn_5 islands had completely transformed. Pure silver was coupled with eutectic Ag-Au-Cu foil in order to study the effect of solutes on interface motion. It was found that baseline shifts and primary solidification led to an underestimation of the remaining fraction of liquid[94]. There was a linear relationship between the interface position and the square root of the isothermal holding-time.

Iron|Tungsten

Joints between tungsten and iron have been obtained (950C, 300s, Ar, 57MPa) by spark plasma sintering, using titanium foil or titanium powder as an interlayer. During thermal cycling, cracks occurred along tungsten|titanium-powder interfaces while tungsten|titanium-foil interfaces remained intact[95].

Nickel|Titanium

When nickel/titanium diffusion couples were prepared by furnace hot-pressing at 650C under a vacuum of 0.1Pa, $NiTi_2$, $NiTi$ and Ni_3Ti formed at the Ni-Ti interface. The Ni_3Ti first grew at the nickel substrate via the diffusion of titanium into the nickel, then $NiTi_2$ appeared and $NiTi$ finally developed in the $NiTi_2|Ni_3Ti$ interface. As the thickness of the diffusion layers increased, part of Ni_3Ti was consumed by the titanium to form $NiTi$. A columnar crystal initially formed within the Ni_3Ti matrix, immediately adjacent to $NiTi$, and then elongated in a direction perpendicular to the plane of the joint. The Ni_3Ti vanished and pure titanium was finally produced. Shear fractures occurred in the Ni_3Ti when samples were prepared at 650C[96]. When prepared at 900C, fracture occurred in the $NiTi_2$.

Self-Bonding of Composites

Metals

Aluminium-Al$_2$O$_3$

The formation of a transient liquid phase bond generally involves a number of common stages: plastic deformation and solid diffusion, dissolution of the interlayer and the base metal, isothermal solidification and homogenization. The joint microstructure depends upon all of those stages. Plastic deformation and solid diffusion promote close contact of the interfaces and liquid-layer formation. In the present case, the microstructure will comprise aluminium solid solution, alumina particles, Al$_2$Cu and MgAl$_2$O$_4$. An important feature can well be alumina particle segregation to the center of the joint. An increase in joint shear strength with increasing bonding temperature is due mainly to an increased fluidity and wettability of the liquid phase and a decrease in the fraction of brittle Al$_2$Cu in the joint. The main reason why a high bonding temperature decreases the joint shear strength is a widening of the alumina-particle segregation region which is a preferred failure site. An increase in joint shear strength with increasing holding time is attributed to a lower fraction of the brittle Al$_2$Cu and to increased homogenization. Particle segregation during the bonding of aluminium-based metal-matrix composites using a copper interlayer was promoted by the slow movement of the solid|liquid interface during isothermal solidification. Alumina particles with diameters of less than 30µm segregated when the copper-foil thickness exceeded 5 to 15µm. When bonding at 853K, the liquid widths which were produced using these copper-foil thicknesses were almost identical to the interparticle spacings in the original materials. When the amount of liquid which formed at the bonding temperature decreased to below a critical level, the material fell apart straight after bonding[97]. The minimum copper-film thickness which was required in order to obtain a sufficiently strong joint increased from 0.6 to 2.4µm when the heating rate to the bonding temperature was decreased from 1 to 0.01K/s. Three situations were considered: one where the liquid width was unrestrained at the bonding temperature, one where the liquid width was kept constant and one where liquid was continually expelled from the bond-line region during joining. Similar experimental and numerically predicted liquid widths were found for an unrestrained liquid width at the start of isothermal solidification. The calculated rate-constant during isothermal solidification was narrower than in reality. When liquid was expelled from the joint interface during bonding at 853K, the liquid width reached a constant value of 55µm. The completion time for isothermal solidification markedly decreased when liquid was expelled from the joint interface, but particulate segregation was still observed in joints which were made using 15- and 25µm-thick copper interlayers[98]. The joint shear-strength increased when the thickness of the copper interlayer was decreased and when the applied load was increase during liquid-expulsion bonding. A higher joint shear-strength resulted from decreased particle segregation in completed joints. The nanocomposite was bonded by using a thin layer of electroplated copper as an interlayer and heating at 580C for 1200s to 1h. With increasing time, isothermal solidification was completed and the final joint microstructure consisted of soft α-aluminium with dispersed CuAl$_2$ precipitates[99]. Decreasing the amount of brittle eutectic in the joint seam, by increasing the bonding time, improved the joint shear strength and, when bonding at 580C for 1h, the highest value was some 85% of the shear strength of the base material. From the fracture surfaces of composite joints, made using a pure copper foil interlayer, it was found that the Al$_2$O$_3$ particles had smooth surfaces, with no adhesions or disruptions. Some areas of the metal at the fracture surface also appeared smooth. This indicated that the particle/metal interface bonding weakened and led to an interfacial de-bonding fracture mode. Bonding using an interlayer which contained both melting-point depressants and active alloying elements was proposed in order to improve the interfacial wettability between weak particle/metal interfaces[100]. When the titanium content was lower, a large number of discontinuous small adhesions was present on the Al$_2$O$_3$ particle surfaces in the joints as a result of reaction, while the particle/metal interfacial bonding became so strong that many Al$_2$O$_3$ particulates were broken during shear testing when the titanium content was high. The composite was also bonded by using Al-Cu and Cu-Ti as interlayers. The particle/metal interfaces were groups into primary (as-prepared) and secondary (after bonding). The shear strength of joints made using copper foil was limited, as noted before, by a weakening of the secondary interfaces, notches around faying surfaces and particle segregation. Bonding using an interlayer containing both melting-point depressant and active elements which could react with the ceramic particles was expected to improve the wettability between the secondary interfaces[101]. This was demonstrated by using a Cu-Ti foil: adhesion formation and broken particles were observed on the joint fracture surfaces; the fracture of the joint partially extended into the sound parent composite. In the case of the Al-Cu interlayer, joints without particle segregation could be obtained, but unbonded regions and voids were observed. As-cast aluminium-30vol%Al$_2$O$_3$ fibre-reinforced composite was bonded by using an Al-12Si and copper interlayer or an Al-12Si-xTi interlayer, where x was 0.1, 0.5 or 1wt%, and heating at 610C for 0.5h under 1 or 0.015MPa in argon. The titanium content and the pressure governed the interfacial wettability and joint-seam microstructure. An improvement in the wettability, due to adding titanium, was confirmed by a reduction in the expulsion of liquid interlayer, the elimination of any interfacial gap, a higher shear strength and a better fracture path. Because there was an incubation period for wetting, reduction of the pressure following melting of the interlayer could further increase the joint shear strength by allowing widening of the remaining seam of solid-solution matrix and by decreasing the fraction of new Al-Si-Ti phase within the seam[102]. A maximum shear strength of 88.6MPa, 99% of the as-cast composite strength, was obtained (table 2) by adding 0.5wt%Ti and using a pressure of 0.015MPa.

Table 2. Effect of the titanium content of an Al-12Si interlayer, and pressure, on the strength of aluminium-Al_2O_3 joints

Interlayer (wt%)	Pressure	Shear Strength (MPa)
0	1	60
0	0.015	75
0.1	1	73
0.1	0.015	80
0.5	1	80
0.5	0.015	89
1	1	50
1	0.015	76

Aluminium-Mg_2Si

As-cast aluminium-Mg_2Si composite containing 15%Mg_2Si was bonded under vacuum by using a copper interlayer. The joints contained α-Al, $CuAl_2$ and Mg_2Si, or α-Al and Mg_2Si, depending upon the bonding temperature and duration[103]. The maximum shear strength was attained upon bonding at 580C for 2h. The microhardness and homogeneity of the joint improved with increasing bonding duration at 560C, while there were no appreciable changes at 580C. The use of a pure copper interlayer habitually causes segregation of the reinforcing particles at the bond interface, and the resultant weakness promotes preferential failure during tensile testing. The use of a mixture of nickel and copper powder leads to less segregation of the reinforcing particles in the central bond zone. The aluminium-15%Mg_2Si composite has also been bonded therefore by using a 1:1 by weight Cu-Ni mixed powder interlayer in argon[104]. As the bonding time was increased, the structural heterogeneity decreased and micro-porosity was eliminated from the central bond zone. The shear strength of the joint increased with increasing bonding time.

Aluminium-SiC$_p$

The composite was bonded by using a copper interlayer and heating at 853K. The strength of joints which were bonded without applying pressure was low due to the formation of metal layers and oxide at the interface. The strength increased with increasing bonding time, and the maximum shear strength was 106MPa; some 48% of that of the composite. The application of 0.2MPa of pressure during bonding greatly increased the joint strength, and a maximum value of 160MPa was obtained; 70% of the composite strength[105]. There was no particle segregation in the interfacial region when the composite contained a high volume fraction of small carbide particles. Because the oxide on the surface of these composites can affect joint properties, sputtering was used to etch the oxide on bonding surfaces by plasma erosion. A 5μm-thick film of Cu|Ni|Cu was then prepared, by magnetron sputtering on the now clean bonding surface, in order to act as an interlayer. A shear strength of 200MPa was obtained; 89.7% of that of the composite[106]. Homogenization of the joint region and an absence of particle segregation in the interfacial region was observed. The composite was also bonded by using a copper film, copper foil, nickel foil or Cu|Ni|Cu multilayer foil as an interlayer. Increasing the holding time could improve the shear strength of the joint. A better strength can be obtained, without any effect of the oxide, by using a copper film as the interlayer and heating at 853K for 2h under 2MPa of pressure. The shear strength of the joints could attain 169.1MPa; some 81.7% of the strength of the composite[107]. The highest strength was obtained by using a Cu|Ni|Cu multilayer foil and heating at 923K for 2h under 2MPa pressure, giving 189.6MPa; some 84.6% of the strength of the composite.

AA1100-Al$_2$O$_3$

A composite comprising 5wt% of alumina particles was bonded by using 5mm of electrodeposited pure copper as an interlayer. Joint formation was attributed to solid-state diffusion of the copper into the aluminium matrix, followed by eutectic formation, with base-metal dissolution and isothermal solidification at the joint interface. Intermetallic phases such as $CuAl_2$ were formed. The concentration of alumina particles increased across the interface with increasing bonding temperature[108]. The highest joint strength of 123MPa was found after heating at 590C for 0.5h.

Materials Research Foundations 43 (2019)
doi: http://dx.doi.org/10.21741/ 9781644900055

AA2024-SiC$_p$

Composite sheet was bonded at 580C by using a mixed slurry of aluminium, copper and titanium powder as an interlayer[109]. A pulsed current density of 1.15×10^2A/mm^2 was then applied, under an original pressure of 0.5MPa, for 0.25 to 1h.

AA2124-SiC

The aluminium metal-matrix composite was bonded by using nickel interlayers, and good bonds were produced with no apparent effect upon the carbide dispersion in the matrix. The absence of segregation was attributed to the high nickel diffusivity in the composite, which led to rapid isothermal solidification at the bonding temperature. Shear tests showed that strengths close to that of the parent metal were possible when thin nickel interlayers were used. Failure occurred at the composite/joint interface[110]. When thick interlayers were used, failure occurred mainly through the centre of the bond-line.

AA6061-Al$_2$O$_3$

The composite was first bonded by using an Al-Cu alloy interlayer. There were no obvious segregated or denuded particulate alumina regions[111]. In another case, the composite was bonded by using Al|Cu|Al or Cu|Al|Cu multi-layer interlayers. After bonding at 600C for 1h by using an Al|Cu|Al multi-layer interlayer, no particle segregation region was evident in the joint and the shear strength was 110MPa[112]. When using a 1.5µmCu|Al|1.5µmCu multi-layer, no particle segregation region was visible in the joint and the shear strength of the joint increased further to 123MPa (figure 11). A composite sample containing 15vol% of Al$_2$O$_3$ particles was bonded using monolithic or nano-particulate interlayers. The formation of eutectic phases such as Al$_3$Ni, Al$_9$FeNi and Ni$_3$Si occurred in the joint zone[113]. It was concluded that the addition of nano-particle reinforcement to the interlayer could improve the joint strength and minimize particle segregation. The 15vol% alumina particle composite was therefore then bonded by using, as an interlayer, electrodeposited nickel coatings containing 18vol% of nanosized alumina particles, with thicknesses ranging from 1 to 13µm[114]. Joint formation was attributed to the solid-state diffusion of nickel into the AA6061 base metal, followed by eutectic formation, base-metal dissolution and isothermal solidification at the joint interface[115]. Intermetallics such as AlFe$_3$Si and Ni$_3$Si formed in the joint zone. The maximum joint strength was equal to 92% of the base-metal shear strength when an 11mm-thick Ni-Al$_2$O$_3$ nanocomposite coating was used as the interlayer[116].

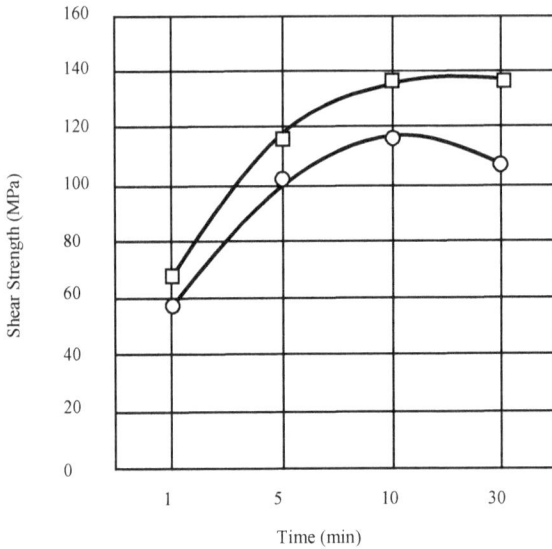

*Figure 11. Effect of bonding time on AA6061-Al₂O₃ joint strength
Squares: Ni-Al₂O₃, circles: Ni coating*

AA6061-SiC

An AA6061-SiC composite was bonded by using a gallium interlayer. The observed kinetics suggested that diffusion occurred along interphase and sub-grain boundaries at below 500C[117]. The composite was also bonded by using a 2:1 mass ratio of aluminium and copper powders as an interlayer. The optimum bonding parameters were 620C and 1h. Many holes were found at joint lines if the bonding was performed at lower temperatures. Oxygen and magnesium were enriched at the joint line. The magnesium in the aluminium alloy reacted with CuO and Al_2O_3 to produce MgO particles and Al_2MgO_4 particles, respectively[118]. This limited the effects of oxide films and improved the mechanical properties of the joints. An extruded 15wt%SiC_p composite was bonded under argon by using a 50μm-thick copper foil as an interlayer and heating at 560C for 20min to 6h under pressures of 0.1 or 0.2MPa. The bonding kinetics were markedly accelerated in the proximity of the reinforcement and this was attributed to an increased solute diffusivity through the defect-rich SiC particle|matrix interface and to porosity[119]. A joint strength that was equal to 90% of the original composite strength was found by bonding under 0.2MPa pressure using a 6h holding time. The high strength was again attributed to the limitation of peripheral oxide at the completion of isothermal solidification, and to the elimination of voids at the joint interface due to solid-state diffusion at a higher (0.2MPa) pressure[120]. Extruded 13vol%SiC$_p$ composite was similarly bonded by using a 50μm-thick copper powder interlayer and heating at 560C under 0.2MPa for times of 20min to 6h. Isothermal solidification and homogenization of the joint region occurred after 3 and 6h, respectively. During isothermal solidification, smaller (11 to 13μm) SiC particles were pushed by the moving solid|liquid interface and segregated around the joint centre-line. Larger (24 to 33μm) particles were engulfed. During isothermal solidification, the larger particles locally hindered solidification-front movement and led to grain refinement[121]. The effective diffusivity of copper in the composite system was about 10^5 times higher than the lattice diffusivity, thus indicating a predominance of short-circuit diffusion through the defect-rich particle|matrix interfaces. When a 15wt%SiC$_p$ composite was bonded at 570C under 0.2MPa using a 200μm-thick copper foil interlayer, the effective diffusivity of copper in the composite system was found to be some 10^6 times higher than the lattice diffusivity[122]. The 15wt%SiC$_p$ was also bonded by using a Cu-Ni powder interlayer and heating at 560C under 0.2MPa for up to 6h. The microstructure of the isothermally solidified zone comprised stable CuAl$_2$ and metastable Al$_9$Ni$_2$ in a matrix of solid-solution together with the reinforcing SiC particles. The microstructure of the central joint zone comprised equilibrium phases such as NiAl$_3$, Al$_7$Cu$_4$Ni and solid solution, together with non-segregated particles and micro-porosity. During shear testing, cracks originated from micro-porosity and propagated along the interphase interfaces, leading to a poor bond strength for shorter holding times. With increasing bonding time, the structural heterogeneity diminished and micro-porosity was eliminated from the central joint zone[123]. After 6h of holding, the microstructure of the central joint zone consisted mainly of NiAl$_3$, with no visible micro-porosity. This gave a joint efficiency of 84%, and failure occurred mainly via decohesion of particle|matrix interfaces. Extruded AA6061-15wt%SiC$_p$ composite was later bonded by using a 50μm-thick mixed Cu-Ag powder interlayer and heating at 560C under 0.2MPa. During an isothermal solidification which required two hours to complete, a ternary liquid phase formed due to the diffusion of copper and silver in aluminium. Subsequent cooling produced a ternary phase mixture, α-Al + CuAl$_2$ + Ag$_2$Al, upon eutectic solidification. When using the present Cu-Ag powder interlayer, isothermal solidification was faster than it was for pure aluminium joints made using a 50μm-thick copper foil interlayer or for composite joints made using a 50μm-thick copper-foil|powder interlayer under similar conditions. The presence of brittle eutectic phases, CuAl$_2$ + Ag$_2$Al, led to a poor joint strength for short bonding times. They disappeared during 2 hours of isothermal solidification, leading to an improved joint strength even when shrinkage had occurred.

Increasing the holding time to 6h eliminated the shrinkage via solid-state diffusion[124]. This produced the highest joint strength of 87MPa; 83% of that of the base material.

AA8090-SiC

Various approaches were proposed for the bonding of aluminium-based metal-matrix composites. One was based upon the simultaneous combination of transient liquid phase bonding and isostatic compression. The bonding of AA8090-SiC composite using this method and a 3μm-thick copper interlayer resulted in bonds having shear strengths of up to 221MPa; 85% of the shear strength of the parent material. Another method combined low-pressure transient liquid-phase bonding under vacuum with isostatic pressing in air. The highest bond strength of a composite bonded using a 7μm-thick copper interlayer was 242MPa; 92% of the parent material strength. In a third method, low-pressure transient liquid phase bonding was combined with hot isostatic pressing[125]. This could be used to fabricate complicated components, with minimal deformation.

A356-SiC$_p$

The A356-SiC$_p$ composite was originally bonded by using a copper interlayer. The joints had a tensile strength that was equal to almost 72% of that of the parent material[126]. As-cast A356-70vol%SiC$_p$ composite was bonded by using an Al-24Mg-16Ga-1wt%Ti interlayer. The idea was to add the titanium, as a melting-point augmenter, to near-eutectic Al-Mg-Ga alloy and produce reinforcement *in situ* by precipitating a fine dispersion of titanium-containing intermetallic reinforcement in the solid-solution joint during isothermal solidification. The bonding was performed by quasi-melting, just full melting or over-melting at 450, 500 and 55C, respectively, under 0.5 to 1.75MPa in argon. The interlayer was able to produce fine (<1μm) dispersed Al-2Mg-1Ga-0.5at%Ti precipitates as *in situ* reinforcement only at 500C under 1 to 1.5MPa. This made the joint-seam microhardness 30HVN greater than that of the matrix. The interlayer exhibited good wettability at every temperature, due to matrix dissolution, and amorphous reaction products containing magnesium and gallium on the SiC surface. Higher bonding temperatures caused a coarsening of needle-like silicon-containing Al-Si-Ti precipitates, while higher pressures produced blocky Al$_3$Ti with a shell of magnesium and gallium. Joints which were produced at 500C under 1.5MPa had a shear strength of 118.6MPa; 98.8% of that of the parent composite. Failure occurred entirely within the composite[127]. When the interlayer alloy, Al-33Cu-6Mg-1wt%Ti, was used the joint shear strength increased with increasing joining temperature from 550 to 600C. When a titanium-free version was used, gaps remained between interfaces, even at 600C. The maximum shear strengths of joints made using Al-33Cu-6Mg-1Ti and Al-33Cu-6Mg interlayers were

62MPa at 600C and 31MPa at 580C, respectively. In the former case, there were partial fracture paths within the A356 matrix while, in the latter case, there was an initial interface fracture path[128]. It was deduced that the titanium improved the wettability between SiC particles and the metallic bond seam and thus increased the shear strength.

Magnesium-SiC$_p$

Magnesium-matrix composites, reinforced with SiC particles, have been bonded by using a Zn-Al-Zn multi-interlayer. A fully solid-solution joint without SiC particle aggregation was obtained by means of ultrasonic-assisted processing at 430C for 30s. The joint shear strength then attained 175.5MPa; 87.5% of that of the base material The time required for isothermal solidification was also shortened to just a few seconds[129], and this was attributed mainly to the ultrasonic vibration having squeezed out a large fraction of the residual liquid phase and facilitated atomic diffusion in the grain boundaries.

Magnesium-TiC$_p$

The TiC-reinforced metal-matrix composite was bonded by using an aluminium interlayer. With increasing bonding time at 460C, the concentration of aluminium in the joint center-line decreased and the joint microstructure changed from α-magnesium solid solution, AlMg and $Al_{12}Mg_{17}$ to α-magnesium, $Al_{12}Mg_{17}$ and TiC particles. At a bonding temperature of 480C, the joint microstructure comprised α-magnesium, $Al_{12}Mg_{17}$ and TiC particles. Titanium carbide particles aggregated along the bond-line when the bonding time was 1h[130]. The elevated fractions of $Al_{12}Mg_{17}$ and the aggregation of TiC particles were the principal factors affecting the mechanical properties. A joint shear strength of better than 58MPa could be attained by heating at 460C or 480C for 20min. The composite was similarly bonded by using a copper interlayer. Upon increasing the bonding time from 5 to 50min at 510C, the average concentration of copper in the joint zone decreased. The joint microstructure changed from copper, α-magnesium and $CuMg_2$ to α-magnesium, $CuMg_2$ and TiC, and the mechanical properties improved[131]. The shear strength of a joint heated at 510C for 50min attained 64MPa due to improved homogeneity. Increasing the bonding time improved the mechanical properties.

Titanium-SiC$_f$

Very good joints were obtained by using a composite interlayer consisting of SiC fibers and Ti-Zr-Cu-Ni alloy[132]. This produced a compact interfacial microstructure of fiber and metal via diffusion.

Ti-6Al-4V-SiC

Somewhere in between the case of a simple alloy and a complex alloy, there is the situation where an essentially simple alloy is also a composite. Continuous SiC-fiber reinforced Ti-6Al-4V composites were joined to a Ti-6Al-4V plate by using Ti-Cu-Zr filler metal. Three processes could be discerned: dissolution of the base metal, isothermal solidification and homogenization. The saturated dissolution width was inversely proportional to the fiber volume fraction. The times required for completion of base metal dissolution, and for isothermal solidification, were expected to lengthen with increasing fiber volume fraction but no marked difference was observed. Homogenization was important in ensuring a high joint strength, and this process could be approximated by means of a one-dimensional diffusion model. The filler metal melted during bonding, and the liquid phase reacted with the SiC fibers leading to the appearance of brittle compounds such as TiC, ZrC and Ti_5Si_3 at the interface between the fibers and the filler metal[133]. They did not affect the joint strength however because they formed only in very small amounts at the ends of fibers. The joint strength increased with bonding time up to 0.5h and attained a maximum value of $850MN/m^2$; 90% of the tensile strength of Ti-6Al-4V. The bonding layer had an acicular microstructure which was composed of Ti_2Cu and α-titanium with dissolved zirconium. Brittle phases such as $(Ti,Zr)_5Si_3$ and $(Ti,Zr)_5Si_4$ formed at the interface between the SiC fibres and the interlayer metal[134]. Those phases existed only at the ends of fibres, and in very small amounts. The joint strength was thus not greatly affected.

Self-Bonding of Simple Alloys

Aluminium-Based

5A02

This alloy was joined by using a two-step process in which the faying area was first heated to 600C for 5s and then cooled to 595C and kept there for 180s. The microstructure and properties were compared with the results of heating at 595C for 180s. The two-step heating could produce a wavy bond line, unlike the planar interface which is associated with conventional bonding at a constant temperature[135]. Defects at the bond line were greatly reduced, and metal/metal contact was established along the wavy bond line by using two-step heating. The tensile strength and bend ductility of the joint were markedly improved.

A multiple-grain phase-field model has been used to explain the formation of such wavy bond-lines during transient liquid-phase bonding. Simulations showed that the principal

Materials Research Forum LLC
doi: http://dx.doi.org/10.21741/ 9781644900055

conditions for unidirectional interface migration could be satisfied by significantly smaller temperature gradients than those predicted by other models[136]. It was also shown that morphological instability could occur not only during solidification but also during melting. In this context, it is interesting to note that the 'father' of morphological instability, Sekerka, performed one of the earliest analyses of transient liquid-phase bonding and outlined the main features of the various bonding stages[137].

AA3003

Bonding of thin foils was carried out by using a silver-nanoparticle interlayer and heating under vacuum to 550 or 570C, at 5 or 25C/min, with a pressure of 1MPa. The nanoparticles aided the formation of joints at below 567C: the aluminium-silver eutectic temperature. The interface contained δ-phase Ag_2Al and μ-phase Ag_3Al. A bond tensile strength of 69.7MPa was obtained by heating at 570C and using a 7μm interlayer[138].

AA6082

The alloy was bonded by using a zincate treatment, followed by copper-plating, to form an interlayer. Isothermal solidification was already complete after 300s. The joint zone contained no eutectic or intermetallic Al-Cu phases. There were however numerous voids of about 30μm in size in the joint zone. It was assumed that these voids formed during bonding due to solidification shrinkage and to the presence of interfacial oxide layers[139]. The average tensile strength was about 270MPa, as compared with a recommended minimum yield strength for the base material of 255MPa. Due to the notch-effect produced by the voids, tensile samples failed with no plastic deformation. The imposition of a temperature gradient during transient liquid-phase bonding produces joints having shear strengths as high as those of the parent material. This method relies upon the formation of non-planar interfaces which are the result of the familiar morphological instability of a solid|liquid interface during solidification of a melt. It is possible to predict the interface morphology and those bonding conditions which produce non-planar bonding surfaces[140]. The prediction was verified experimentally while bonding pure aluminium, and also the present alloy, by using copper interlayers.

AA7075

he alloy was bonded by using a liquid gallium interlayer and heating at 460C for 600s under 10MPa of pressure before homogenization (465C, 2h) and T6 heat-treatment. The fatigue life of the base alloy was 10^7 cycles under 90MPa, while the fatigue life of bonded material at the same stress amplitude was 1.2×10^6 cycles; some 10% of the total alloy fatigue life[141].

AA7475

This superplastic alloy was bonded by using 20 or 10μm-thick electroplated zinc interlayers and heating at 515 or 530C under uniaxial pressures of 1, 2 or 3MPa for 0.5 to 6h. Overlap shear tests showed that, for suitable bonding conditions and surface pre-treatments, lap shear strengths (in the T6 condition) which were greater than 90% of that of the parent alloy could be attained[142].

AA8090

The alloy was bonded by using copper interlayers. The transient liquid phase formation and melt solidification were governed by the solid-state diffusion of copper along Al-Li alloy grain boundaries[143]. The shear strengths of the joints were comparable to those of solid-state joints and were greater than 90% of that of the base metal.

Al-Mg

A one-dimensional mathematical model was used to predict re-melting and re-solidification during the transient liquid phase bonding of Al-Mg alloy by taking account of macroscopic solute diffusion in the liquid and solid, together with solid transformation into liquid caused by solute macro-segregation[144]. It was shown numerically that the holding-time and holding-temperature strongly affected the solute distribution, and that in turn markedly affected the mushy-zone thickness. The one-dimensional model was similarly used to predict the solute redistribution during the bonding of Al-Cu alloy and unsurprisingly led to the same conclusion[145].

Al-Si

When Al-Si alloy was bonded by using a copper interlayer, the copper reacted with the aluminium of the alloy to form a liquid eutectic liquid phase, while the silicon from the alloy tended to prevent the reaction. The microstructure of the joint comprised α-Al, Si and intermetallics such as $CuAl_2$ and Al_4Cu_9. The fraction of intermetallics decreased with increasing bonding time. With increasing bonding time, the shear strength of the joint first increased and then declined. A shear strength of 70.2MPa could be obtained by bonding at 560C for 2h. Fracture occurred at the bonding-region|base-metal interface during shear-strength testing, and there was a change from brittle, to brittle/ductile, fracture with increasing bonding time[146]. The alloys, Al-27Si and Al-50Si, were bonded by using a thin copper interlayer. Joints free from obvious defects were obtained and a wide seam comprising fine eutectic and coarse silicon particles formed at the Al-27Si|Cu|Al-50Si joint[147]. Introduction of silicon into the liquid phases during solidification explained the formation of the larger silicon particles and ultra-small silicon

Materials Research Forum LLC
doi: http://dx.doi.org/10.21741/ 9781644900055

particles in the seam. The shear strength of the joint could attain 63MPa. Cast Al-7Si-0.3wt%Mg alloy has been bonded by using a Sn-20Cu-5wt%Ge interlayer. The interlayer material comprised beta-tin, η-phase (Cu_6Sn_5) plus germanium-rich precipitates. The joint had a much more complex microstructure which consisted mainly of beta-tin plus various aluminium-copper and aluminium-germanium phases[148]. Small silicon-oxide rods and an hexagonal aluminium-copper-magnesium-germanium phase having the lattice parameters, a = 0.7123nm and c = 2.40nm, was found in the seam of the joint. Rapid ultrasound-induced transient liquid phase bonding has been used to join Al-50Si alloy, with zinc foil as an interlayer, at 390C in air. This temperature was below the melting-point of the interlayer. Fracture of oxide films along the edge of silicon particles led to contact and interdiffusion between the aluminium component and the zinc interlayer; producing liquid Zn-Al alloy. The width of the Zn-Al alloy gradually decreased with increasing ultrasonic vibration time. This was due to liquid being squeezed out, and to the acceleration of diffusion. During this isothermal solidification stage, the completion time was greatly shortened. Acoustic streaming and ultrasonic cavitation were induced in the liquid metal. With passing time, more and more silicon particles migrated to the center of the soldered seam. The highest average shear strength of the joint was 94.2MPa, and fracture occurred mainly in the base metal[149].

Copper-Based

Cu-Be

Copper-beryllium alloy can be bonded by using silver-based interlayers and heating under vacuum or in argon for various holding times. A maximum tensile strength of 156.45MPa was obtained by heating at 780C. The fatigue strength of a joint made under vacuum was higher than that of one made in argon[150]. Brittle phases formed in the joint region, markedly reducing the joint strength and toughness. This was especially true when using cadmium-containing interlayers, and $CdCu_2$, $CdCu_4$, Ag_5Zn_8, CuZn, Cu_3Sn and Ag_3Sn were found in the fusion zone[151]. Diffusion of the principal elements from the interlayers and into the base metal was the main factor governing microstructural evolution of the joint interface. Alloying elements diffused from the interlayers and into the alloy grain boundaries.

Cu-Cr-Zr-Ti

This alloy was bonded by using thin electroplated coatings of nickel and manganese. One part was given a 4μm nickel coating followed by a 15μm manganese coating, while the other part was given only a 4μm nickel coating. The assembly was heated at 1030C for 0.25h, furnace-cooled to 880C and held for 2h followed by argon-quenching to room temperature. The interlayer underwent isothermal solidification during the 0.25h spent at 1030C. The interlayer had a final composition of Cu-17Ni-9wt%Mn at the joint center, with a steep decline in the nickel and manganese concentrations in going from the joint center to the base metal interfaces. During holding for 2h at 880C, these gradients flattened out appreciably and the joint acquired an homogeneous composition, of Cu-11Ni-7wt%Mn, at the its center[152]. During lap shear-tests, failure occurred in the base metal at points away from the joint region. The base metal underwent appreciable grain-coarsening due to the high-temperature exposure during bonding and thus exhibited a marked reduction in yield strength.

Iron-Based

Iron-based, nickel-based and commercially-pure copper interlayers are habitually used to bond low-carbon steel, which is then heat-treated to produce dual-phase steel. All three interlayers produce joints having desirable mechanical properties, but produce differing microstructures in the joint region and require different periods for completion of the bonding process. Bonding in air also produces similar microstructures and mechanical properties to those produced by bonding under vacuum.

Fe-C

Carbon-steel tubes were bonded by using pure copper interlayers and heating at 1300C for 420s under a pressure of 5MPa[153]. Whereas the use of an amorphous Fe-B-Si interlayer[154] led to completion of the bonding over the entire bond area, bonding using a copper interlayer led to only partial completion. Because copper is a cementite-promoter, the amount of cementite which coexisted with ferrite grains was higher in the joint region. The opposite effect was observed when using Fe-B-Si interlayers because cementite cannot form in silicon-enriched zones and only ferrite grains were then present at the joint[155]. The joined tubes failed away from the bond in the heat-affected zone in both cases and attained 96% of the ultimate tensile strength of the base material, in the as-bonded condition[156]. Earlier bonding methods had used Fe-3.8wt%B amorphous ribbons as an interlayer, with heating to 1250C under pressures of 2, 3 or 4MPa. Bonding was complete in 7min when a pressure of 4MPa was used, but remained incomplete at lower pressures[157]. Plastic deformation at the joint was observed. As the pressure which was used increased, the amount of pro-eutectoid ferrite decreased and there was a consequent increase in the hardenability of the steel which was attributed to the effect of compressive plastic deformation at the joint[158]. A numerical model for the isothermal solidification time was developed for bonding under plastic deformation and was applied to carbon-steel sandwich panels that were bonded by using a copper interlayer. A reasonable isothermal solidification time was predicted when an effective diffusion coefficient was used, and that time was clearly shortened by the plastic deformation[159,160]. Diffusion bonding of high-carbon steel has been performed by vacuum brazing (900 to 1050C, 0.5h) under an uniaxial load, using a nickel foil interlayer. The microstructure of the steel changed from martensitic to austenitic at the bonding temperature, and then to ferritic-pearlitic during cooling to room temperature. The diffusion zone did not contain any intermetallic compounds. The bond strength was determined by the degree of solid-solution hardening and the contact-area of the surfaces. A maximum ultimate tensile strength of about 532MPa and a shear strength of about 792MPa were found for a joint processed at 1050C; higher than the reported values for martensitic steel[161]. Dissimilar steel pipes have been joined by using a transient liquid phase two-step heating process under argon and a FeNiCrSiB amorphous foil. Solidification of the liquid interlayer occurred on the substrate, and the grains grew in a specific direction. A curved interfacial layer was formed because a small amount of impurities collected at the center of the joint[162]. Short-term high-temperature heating could promote melting of the substrate and ensure epitaxial growth of the grains.

20 Steel

Pipes have been bonded by using nickel-based BNi2 or an iron-nickel based foil as the interlayer. The BNi2 interlayer produced nickel solid-solution joints which contained silicide precipitates and had an heterogeneous composition. Bonding using the iron-nickel based foil resulted in homogenous joints which had microstructures and compositions that were similar to those of the parent metal and which were free from silicide precipitates[163]. The mechanical properties of these joints were better than those of BNi2-bonded joints; in particular, the toughness was close to that of the parent metal.

Fe-Cr-W

Oxide-dispersion strengthened martensitic steel has been bonded by using an oxide-dispersed interlayer, Fe-9Cr-2W-0.2Ti-0.5C-3B-2Si-0.35Y_2O_3, which was prepared by means of mechanical alloying and spark-plasma sintering[164]. As compared with conventional transient liquid phase bonding using an oxide-free interlayer, Fe-3B-2Si-0.5C, the microstructure of the molten zone contained finer grains in the joint when oxide was present, and the grain size was some 67% smaller. This then increased the hardness by 100VHN in the region of the joint. The grain-refinement was attributed to an increased nucleation at oxide particles. The bonding 9%Cr oxide-dispersion strengthened steels, again using the Fe-3B-2Si-0.5C interlayer, was carried out by heating at 1180C for

0.5 to 4h. The precipitation of chromium boride, which was a feature of the 19%Cr oxide-dispersion strengthened steel, was avoided in the present steel due to the lower chromium content. The silicon content tended to be slightly higher within the bonding zone[165]. The agglomeration and coarsening of Y_2O_3 particles in the 9%Cr steel led to softening of the bonding zone formed by incipient melting of the foil interlayer. In another bonding technique, a thin pure-boron interlayer was provided by means of electron-beam physical vapor deposition in order to join the oxide-dispersion strengthened steel, Fe-15Cr-2W-0.2Ti-0.35Y$_2$O$_3$, via uniaxial hot-pressing. The joint was free from voids at the bonding interface[166]. The tensile strength of the joint was similar to that of the material itself, but the total elongation decreased at 700C and was associated with failure at the joint interface[167]. The latter was attributed to partially discontinuous microstructures which were aligned along the interface.

Fe-Mn-Mo

The geological drill pipe steel, 45MnMoB, was joined by using the two-step heating process of 60s at 1250C and 120s at 1230C[168]. This produced an homogeneous joint exhibiting a microstructure and composition which were equivalent to those of the metal itself; with a tensile strength of 890MPa and permitting bending through 180°. In a development of the method, other two-step processes were used. The bonding area was first heated to 1250C, kept there for 5s and then cooled to 1210, 1220 or 1230C and kept there for 120s. This could produce a cellular interface, during the first stage, which was different to the planar interfaces normally expected. No bond interface could be observed in the final joint, and the microstructure of the joint was similar to that of the metal itself. When the second-stage temperature was increased, the presence of oxide-scale particles and porosity in the joint region diminished in terms of both size and number and the bond strength increased[169]. The joint failed through the base metal, and no failure occurred when the joint was pulled or bent through 180° at 1250 and 1230C, respectively. This steel was also bonded under a low imposed pressure by using an iron-based interlayer; the maximum deformation caused by the pressure being 1%. Applying the low pressure during bonding could produce a so-called ideal joint, with no joint interface and identical in composition to that of the metal itself, within 180s. Insoluble oxide particles were observed in joints made using a 5C/cm temperature gradient, and the joint tensile strength was 75% of that of the metal itself[170]. On the other hand, few oxide particles or impurities were found in joints made using a 50C/cm temperature gradient, and the joint failed through the base metal.

Materials Research Forum LLC
doi: http://dx.doi.org/10.21741/ 9781644900055

Fe-Ni-Cr

The Fe-35Ni-26wt%Cr alloy was bonded by using AM17 or MBF-50 interlayers containing boron, silicon and chromium. The melting point and critical interlayer width decreased with an increase in the concentrations of those elements, and the optimum levels were about 3%B, 4%Si and 3%Cr[171]. For a bonding temperature of over 1423K, the tensile (figure 12) and creep-rupture strengths of the joints were of the same magnitude as those of the metal itself.

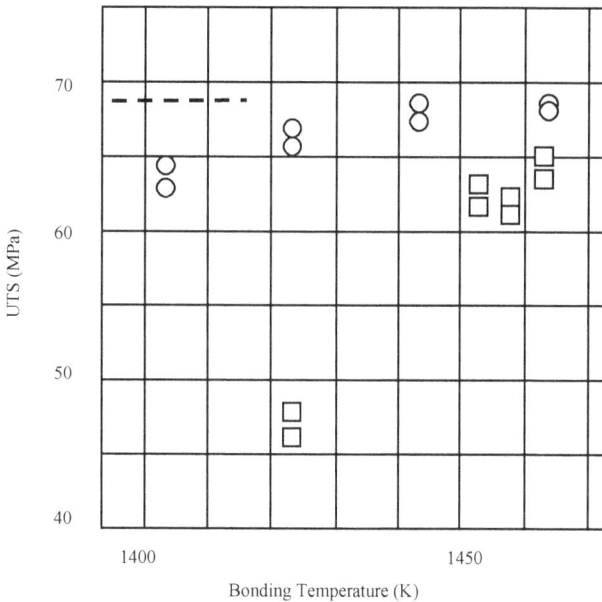

Figure 12. Effect of bonding temperature on
the tensile strength of Fe-35Ni-26wt%Cr joints
Circles: AM17 interlayer, squares: MBF-50 interlayer
Dotted line indicates the strength of the material itself

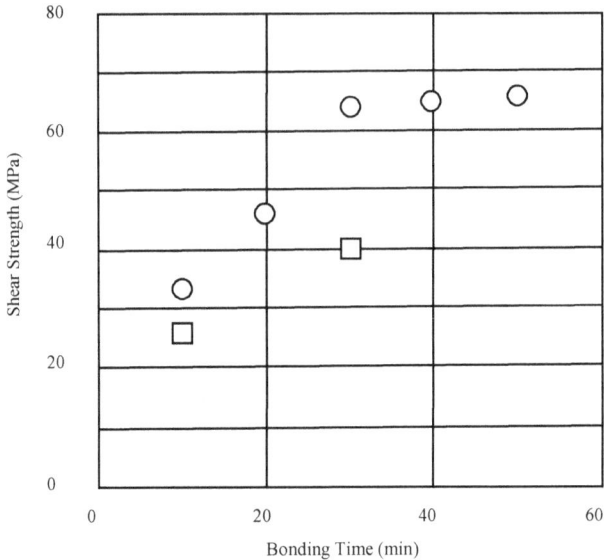

Figure 13. Shear strength of AZ31 bonds as a function of bonding time
Circles: copper\tin interlayer. Squares: copper interlayer

T91

Pipes made from this material have been joined by using FeNiSiB amorphous alloy as an interlayer and heating in argon. The solidified joint comprised a primary solid solution having a composition similar to that of the parent metal, and was free from precipitates[172]. Heating at 1250C for 180s under a pressure of 6MPa produced the optimum result. Bonding by heating at 1270C for 30s and at 1230C for 180s produced joints having tensile strengths which were as high as those of the parent material[173]. The joint strength was possibly enhanced by an improved metal-to-metal contact along the non-planar bonding lines. In a further method, pipes were joined by using BNi2, Fe78Si9B13 or FeNiCrSiB amorphous interlayers in argon. The tensile strength and bend strength at room temperature of the joints made using a Fe-Ni-Cr-Si-B amorphous interlayer were equal to, or greater than, that of the material itself. Borides were found in the joint region when bonding was performed using the Fe78Si9B13 interlayer. Use of the nickel-based BNi2 foil produced a bond region which was stabilized by the high nickel concentration[174]. Fracture of the joints was attributed to the presence of brittle intermetallics at the interface, and to the discontinuity of the bond line microstructure.

Magnesium-Based

AZ31

The alloy was joined by using copper coatings, or copper coatings plus a tin interlayer. The bonding temperature was 520C in all cases, while various bonding times were used. The joints which were produced by using only copper coatings were weaker as compared with those made using a copper coating plus a tin interlayer (figure 13). Particles of Cu_2Mg were found in the joint in all cases, but the joint was predominantly a solid solution which was rich in magnesium. The tin interlayer did not contribute to the intermetallics in the joint and diffused away through the magnesium matrix. Within the joint interface, there was a slight increase in microhardness as compared to that of the base metal.

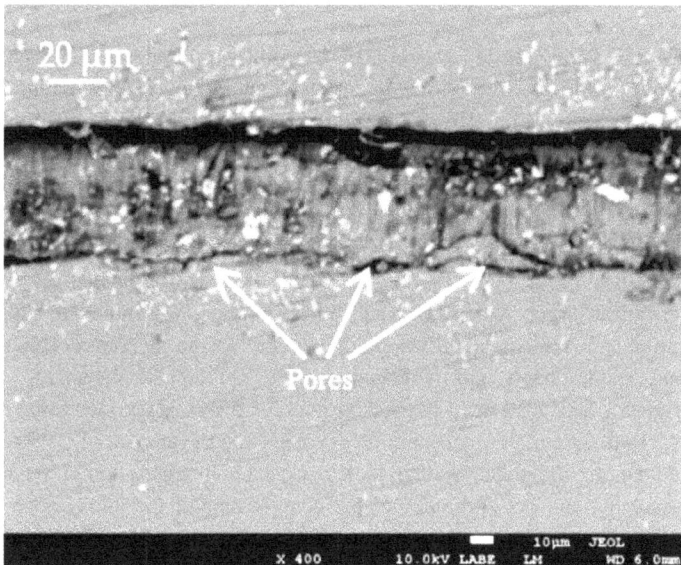

Figure 14. Scanning electron microscopic back-scattered images of a joint made in AZ31 alloy using a copper coating and bonding for 30min

This was attributed to the formation of intermetallics in the joint[175]. The use of a tin interlayer improved the joint strength by reducing the numbers of pores in the joint. Such pores were clearly observed (figure 14) in joints made using copper coatings, especially where fracture had occurred. In another case, the alloy was coated with 5μm of nickel before bonding at 520C under 8MPa of uniaxial pressure at 300s to 1h. The nickel coating could not be detected after bonding at 0.5 or 1h, because of its complete dissolution. Second-phase particles which contained Mg_2Ni were observed near to the joint region[176]. The shear strength of the joint first increased, with increasing bonding time, up to 20min and then decreased. When bonding was performed using a pure nickel interlayer, an eutectic formed between the magnesium and nickel during heating at 515C for 300s to 2h. As the bonding time was increased to 1h, the eutectic zone completely disappeared and the joint solidified isothermally. This gave a maximum hardness of 179VHN and the highest joint shear strength of 36MPa. When the bonding time was increased to 2h, the shear strength of the interface decreased, presumably due to the formation and segregation of brittle Mg-Ni intermetallics within the joint[177].

Mg-Al-Zn

When the Mg-3Al-1Zn alloy was bonded by using an aluminium interlayer, the joint composition profile and microstructure depended upon the bonding time at 480C. Upon increasing the bonding time from 1 to 60min, the concentration of aluminium and the amount of $Al_{12}Mg_{17}$ in the joint decreased. Following heating at 2h, the most marked features of the joint were homogenization of the composition and grain coarsening. The brittle $Al_{12}Mg_{17}$ and grain coarsening were the main causes of shear-strength impairment[178]. A shear strength of 76.1MPa, 92.4% of that of the metal itself, was obtained by heating at 480C for 1h. The alloy could also be bonded by using copper interlayers in an argon atmosphere, and joint formation was found to involve plastic deformation and solid-state diffusion, dissolution of the interlayer and base metal, isothermal solidification and homogenisation. The joint composition profiles and microstructures depended upon the bonding time at 530C. Upon increasing the bonding time from 10 to 60min, the concentration of copper and the amount of $CuMg_2$ in the joint decreased. During prolonged heating, there occurred homogenisation of the composition, grain-coarsening and disappearance of the bonding line at the joint centre. The brittle $CuMg_2$ phase and the grain coarsening were the main cause of an impaired joint shear strength[179]. A joint shear strength of 70.2MPa, 85.2% of that of the metal itself, was obtained by bonding at 530C for 0.5h.

Mg-Mn-Ce-Zn

The alloy was bonded by using a brass interlayer and applying ultrasonics at 460C. With increasing ultrasonic time, the brass interlayer gradually disappeared and Mg-Cu-Zn eutectic formed. The amount of eutectic in the joint decreased as the ultrasonic time was further increased. A continuous oxide film at the interface was partially fractured by the ultrasonic vibration, and was then suspended over melt due to the eutectic reaction. The suspended oxide film then broke into small fragments due to ultrasonic cavitation and these were finally squeezed out of the joint by an ultrasonic squeezing action[180]. The maximum shear strength of the joint attained 105MPa (figure 15); equal to that of the base metal.

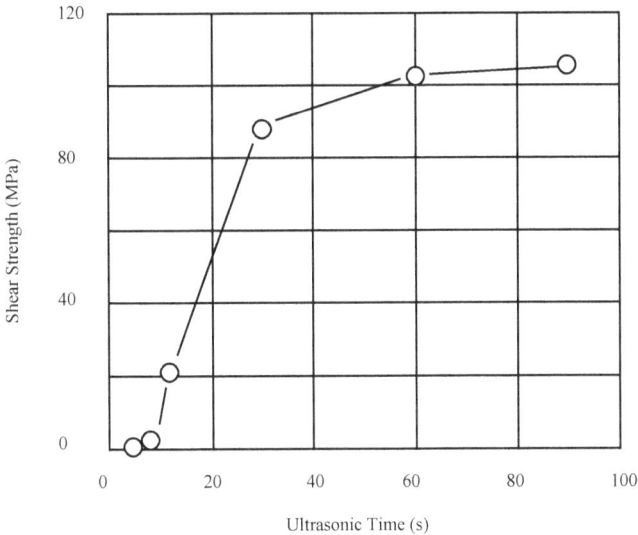

Figure 15. Shear strength of ME20M alloy joints as a function of ultrasonic exposure time

Nickel-Based

Ni-Al

The bonding of a Ni-10.3at%Al alloy was carried out by using a Ni-10at%B interlayer and the results were simulated using finite-difference methods. A thermodynamic assessment of the Ni-Al-B system was used to determine the phase diagram and the thermodynamic factors of the diffusion coefficients. Composition-dependent diffusion mobilities were deduced for the ternary system. The predicted melt-widths as a function of time agreed well with experiment. The calculated and experimental aluminium profiles agreed with regard to the matrix but not with regard to the liquid[181]. The simulations predicted the observed precipitation and dissolution of $Ni_{20}Al_3B_6$ intermetallic in the base material.

NiAl

Microstructure development during the transient liquid phase bonding of cast polycrystalline near-stoichiometric NiAl was investigated early in the use of the method. Bonds were made by using BNi-3 (Ni-4.5Si-3.2wt%B) interlayers in the form of melt-spun foils and heating at 1065 or 1150C. Transformation of the substrate β-phase into an L10-type martensite, due to aluminium transfer to the joint, was reported[182]. The formation of L12-type (γ′-structure) layers at the joint|substrate interfaces was also considered, as was the formation of $M_{23}X_6$-type borides within the joint and the surrounding substrate. Isothermal solidification in NiAl|Cu|Ni joints, and the effect of copper diffusion from the joint centerline upon the microstructures of the adjacent NiAl and nickel substrates was investigated. Changes in the microstructure of the bond center-line due to the entry of the aluminium from the NiAl substrate and of nickel from both the NiAl and nickel substrates were further considered[183].

Ni-Co-Cr

The microstructures which develop during the bonding of dissimilar superalloys are more complex than those which develop between similar materials[184]. A general study was made of the isothermal solidification stage during the bonding of monocrystalline superalloys, using Ni-7.5Co-6.0Cr-5.9Al-5.8W-1.2Mo-1.1wt%Ti as an example. This revealed that the isothermal solidification stage deviated from standard parabolic transient liquid-phase models. Many borides having fine short rod and acicular morphologies formed in the diffusion-affected zone of thick-wall and thin-wall substrates at the isothermal solidification stage. Small dendritic precipitates were identified as being M_3B_2 borides and intergrowths of M_3B_2 and M_5B_3 borides[185]. A boron composition peak existed in the diffusion-affected zone. There were three stages of isothermal solidification: an initial stage, a transient stage and a final stage, each involving differing growth velocities of the isothermal solidification zone[186]. The dependence of the width of the isothermal solidification zone upon the square root of the bonding time did not satisfy the parabolic relationship. Samples having various misorientations were bonded, and creep-tested at 760C under a stress of 780MPa. The creep life decreased with increasing relative misorientation (figure 16). Failure occurred in the bond region, but depended upon that misorientation. In low-angle boundary specimens, cleavage began at defects and grew perpendicularly to the tensile stress before connecting via the various slip planes around the cleavage planes. In this case, deformation involved dislocations and stacking faults on multiple planes. With increasing misorientation, the dislocations and stacking faults moved on a single plane[187]. The dislocations thus interacted with the grain boundary and led to fracture. It was concluded that the orientations of bonded superalloy monocrystals should be arranged so that the grain boundaries thus introduced are relatively small.

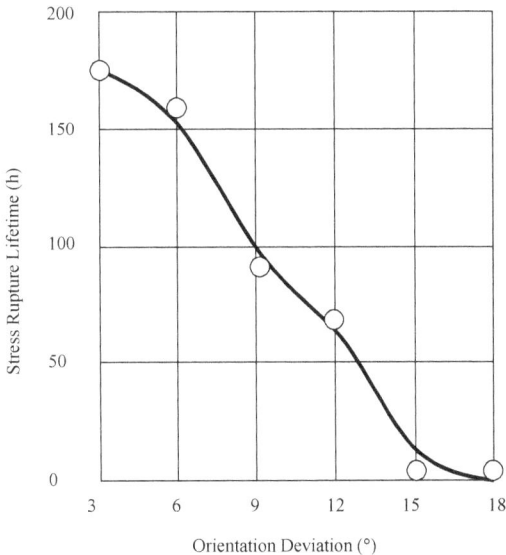

Figure 16. Stress-rupture lifetime of Ni-Co-Cr alloy
as a function of orientation deviation

Ni-Fe-Nb

The effect of the interlayer composition upon the solid-solution strengthening of a bonded cast Ni-Fe-Nb superalloy was studied for the interlayers: Ni-4.5Si-3.2B, Ni-7Cr-4.5Si-3.2B-3Fe and Ni-15.2Cr-4B. The hardness of the isothermal solidification zone was the factor which controlled the shear strength of the joint. It was concluded that the initial chromium content of the interlayer, and the extent of base-metal dissolution (in turn controlled by the initial boron content) played important roles in strengthening the isothermal solidification zone[188]. Use of the Ni-15.2Cr-4B interlayer led to an eutectic-free bond possessing the highest shear strength.

Titanium-Based

TiAl

Bonding has been performed by using titanium foils, combined with copper, nickel or iron, as an interlayer. Although the interface structures of the joints were different, they all comprised five sub-layers: two diffusion zones, two interfacial zones and an interlayer. Formation of the transient liquid phase controlled the diffusion behavior of the melting-point depressing copper, nickel and iron atoms, leading to the formation of the various interface structures of the joints[189]. Post-bonding heat treatment had no obvious effect upon the interface structure of the joints because of the low diffusion coefficient of copper, nickel and iron in the parent material[190]. There was a slight resultant decrease of the microhardness in the joint area. When bonding was performed using a Ti-Ni interlayer, a continuous α_2 layer formed at the joint interface when the bonding temperature was less than 1125C. The α_2 layer hindered interdiffusion between the TiAl and the liquid interlayer, resulting in a long period up to complete isothermal solidification. When the bonding temperature was 1150C, no continuous α_2 layer formed, and the isothermal solidification rate was higher. With decreasing nickel content in the interlayer, the isothermal solidification rate was reduced due to a decrease in the dissolution of TiAl in the liquid interlayer[191]. The highest shear strengths at room temperature and at 800C were 281MPa and 243MPa, respectively, when bonding was performed at 1150C for 300s. When bonding was carried out using a Cu-Zr-Al amorphous alloy foil as an interlayer, defect-free joints were obtained. Diffusion reactions occurred between the TiAl and the interlayer, leading to the formation of three reaction layers, plus some intermetallic phases[192]. The highest room-temperature shear strength of 80.4MPa was obtained by heating at 900C for 0.5h. The γ-TiAl alloy known as Gamma Met PX was bonded by using a composite interlayer which consisted of a non-melting component (TiAl alloy) and a liquid-forming component (copper)[193]. The joints could retain room-temperature mechanical properties that were better than 90% of those of the parent material[194].

Ti₃Al

This material could be bonded by using a TiZrNiCu interlayer. With increasing bonding temperature and time, the width of the bonding zone increased and that of reaction zones decreased[195]. When the bonding time was 1h at 900C, the joint microstructure consisted of titanium solid solution, Ti₃Al and Ti₂Cu. A joint strength of 420MPa could be obtained. When pure copper was used as the interlayer, the latter dissolved slowly at the bonding temperature of 900C. With increasing bonding temperature and time, isothermal solidification of liquid in the joint began and the liquid width decreased. A higher bonding temperature and longer time produced an homogeneous composition and microstructure. The width of the bonding zone increased meanwhile, and the number of reaction zones decreased. The joint comprised mainly a solid layer, a liquid layer and solid-liquid layer at 900C. When the temperature and time were further increased, the joint comprised mainly a solid layer and a residual liquid layer[196]. A joint shear strength of 307.5MPa could be obtained by heating at 950C for 1h.

Ti-Al-Cr-Nb

The alloy, Ti-48Al-2Cr-2at%Nb, was bonded using wide-gap composite interlayers consisting of copper as a liquid former plus a Ti-Al alloy as a non-melting diffusional sink[197]. Other interlayers essayed were 5 or 50μm-thick commercial-purity copper foils. The formation of copper-containing intermetallics at the joint line and the formation of γ-TiAl by epitaxial growth and non-epitaxial precipitation were noted[198]. The process via which copper wets the alloy was identified and was suggested to be quite different to that occurring during other bonding processes. The joint region had the same microstructure as that of the base alloy, but contained small amounts of copper[199]. The oxidation behavior of the alloy showed that this copper was not detrimental to the oxidation resistance of the alloy[200,201].

Ti₂AlNb

When Ti₂AlNb was bonded by using a Ti|Ni interlayer, the joint consisted mainly of 2 zones: an isothermal solidification zone and a thermal solidification zone. The typical interfacial microstructure was: Ti₂AlNb|B2|Nb₃Al+B2+τ3+Ti₂Ni|Ti₂AlNb. With increasing time, the Nb₃Al and Ti₂Ni vanished, the τ3 phase gradually decreased and the B2 phase increased[202]. The maximum room-temperature shear strength could reach 428MPa when the joint was made by heating at 1180C for 1200s. The strength was 423,

Materials Research Forum LLC

doi: http://dx.doi.org/10.21741/9781644900055

407 and 338MPa for 500, 650 and 800C, respectively. Cracks propagated mainly through the τ3 phase in the thermal solidification zone[203]. When it was bonded by using a Ti-Cu-Zr foil interlayer and heating at 950, 1000 or 1050C, the joint consisted of an isothermally solidified zone and a diffusion-affected zone. A non-isothermally solidified zone existed only for lower bonding temperatures. The interface morphology of the joint was highly dependent upon the bonding temperature. When that temperature was increased from 950 to 1000C, the width of the non-isothermally solidified zone decreased from 69 to 23μm. When the bonding temperature was 1050C, the non-isothermally solidified zone disappeared. With increasing bonding temperature, the constitution of the joint changed from $Ti(Cu,Al)_2$ + $AlNb_2$ + titanium solid-solution to titanium solid solution + $Nb(CuAl)$ + Al_4Cu_9 + Al_2Zr_3, and the proportions of the secondary phases were 35.7, 20.2 and 6.7%, respectively[204]. The morphology changed essentially because high bonding temperatures were above the 980C α→β transition temperature. When an amorphous Ti-Cu-Zr-Ni foil interlayer having a melting point of 1133K was used and bonding was performed at 1153 to 1223K for 600 to 3000s, the joint strength was greatly affected by the reaction-layer thickness. The optimum bonding conditions were 1173K and 600s. The maximum tensile strength of the joint attained 260MPa. The activation energy and growth velocity of the reaction layer in the joint were calculated to be 161.7kJ/mol and $0.21m^2/s$, respectively[205].

Ti-6Al-4V

Transient liquid phase bonding of Ti-6Al-4V, using a copper interlayer, was investigated by varying the bonding time at 930C. Intense metallurgical reactions occurred at the Ti-6Al-4V|Cu interface and a joint consisting of TiCu, Ti_2Cu layers and an eutectoid structure formed during a bonding time of 60s. Prolonged holding led to transformation of the TiCu into Ti_2Cu, with progressive subsequent decomposition of the latter. An homogeneous joint consisting mainly of α-Ti matrix containing small Ti_2Cu platelets was eventually obtained by bonding for 0.5h. The homogenized joint had mechanical properties which were comparable to those of the Ti-6Al-4V[206].

Ti-Nb-Al

When Ti-25Nb-22at%Al alloy was bonded by using a Ti-15Cu-15wt%Ni interlayer it was found that a suitably long time and high temperature favoured the formation of joints having an homogeneous composition and high strength. Niobium was the main element controlling the formation of a compositionally homogeneous bonding zone. The latter, following rapid cooling from the bonding temperature, comprised B2 phase. Slow cooling tended to improve the joint strength[207]. The room-temperature tensile strength of a joint made by heating at 970C for 1.5h, with slow cooling, attained 1018MPa; 93% of that of the material itself. The strength was 931MPa after rapid cooling.

Ti-Ni-Cu

The Ti45Ni49Cu6 was bonded by using a copper interlayer and heating in argon under a pressure of 20MPa; just below that which would cause macro-deformation. Mutual copper, titanium and nickel atom migration was detected along the bond interface[208]. Good joints were obtained by heating at 940 to 970C for 40 to 60min.

Self-Bonding of Complex Alloys

Cobalt-Based

FSX-414

The solutionized and aged cast superalloy could be bonded by using an MBF-30 interlayer. Due to complete isothermal solidification, no eutectic structure was observed in the joint region but some CoB intermetallics were present in the diffusion-affected zone. Homogenized joints exhibited a more uniform distribution of the alloying elements, as compared with inhomogenized ones. The hardness of homogenized specimens was closer to that of the base-metal hardness, as compared with non-homogenized specimens. A hardness-peak occurred in the diffusion-affected zone of both homogenized and non-homogenized joints, and the shear strength and shear fracture energy of the homogenized joint were equal to some 93% and 82% of those of the base metal, respectively[209]. More extensive fibrous zones and smaller dimples were observed on the fracture surfaces of homogenized joints, as compared with non-inhomogenized ones. When the superalloy was bonded by using an MBF-80 interlayer, a bonding time of 30s to 0.5h and a gap-size of 25 to 100µm, continuous center-line eutectic phases comprising nickel-rich and chromium-rich borides were found in the joints following incomplete isothermal solidification. Globular and acicular phases were present in the diffusion-affected zone, and were thought to have been nickel-chromium or cobalt-chromium borides. The time which was required for complete isothermal solidification increased with increasing gap-size. This was consistent with calculations which were based upon the assumption of diffusion-induced solid|liquid interface movement. A deviation from these calculations was however noted for 75 and 100µm gaps[210]. Under complete isothermal solidification conditions, the shear strength and hardness of the isothermal solidification zone decreased with increasing gap size and the fracture surfaces of specimens suffering incomplete isothermal solidification exhibited secondary crack paths through the brittle center-line eutectic constituents.

K640

This superalloy could be bonded by using short-term high-temperature heating to melt the interlayer, followed by isothermal solidification of the liquid phase at a lower temperature. This two-step technique could reliably produce a good joint of uniform chemical composition and having better mechanical properties than those bonds produced by using a single-step process[211]. The optimum treatment in the present case was to heat at 1250C for 0.5h and then at 1180C for 3h. The high-temperature tensile strength then attained 74% of that of the base material, as compared with 58% for the single-step process.

Iron-Based

2205

This duplex stainless steel could be bonded by using Ni-B-Si or Fe-B-Si amorphous interlayers: the nickel-based interlayer stabilised the austenitic phase along the bond length and consequently hindered grain growth across the joint region, while the iron-based interlayer produced bonds which were homogeneous, with regard to microstructure and composition across the joint. The mechanical and pitting-corrosion properties were similar to those of the alloy itself[212]. Bonding could also be accomplished by using a copper interlayer. Joints were produced by isothermal solidification, while rapid heating and cooling cycles suppressed the formation of σ-phase in the joint. An homogeneous microstructure, with grain growth and a good α/γ phase balance across the bond-line, was obtained. This again imparted mechanical and corrosion-resistant properties which were similar to those of the alloy itself. Analytical models for the isothermal solidification and homogenisation stages when bonding using a copper interlayer could be used to estimate the time required for the completion of those stages by taking account of the lattice and grain-boundary diffusivities of copper in iron[213]. Good agreement was found, between the predicted time required for isothermal solidification and experimental data, when an effective diffusivity was used which was the geometric mean of the lattice and grain-boundary diffusivities in α-ferrite. This indicated that lattice and grain-boundary diffusion through the α-phase played an important role in bonding using a copper interlayer[214]. On the other hand, the predictions for the homogenisation stage deviated from experimental observations when the copper concentration in the joint region approached a value which was close to that of the base metal.

A286

Study of the bonding of the iron-based heat-resistant stainless steel showed that it was more difficult when using a nickel-based interlayer that contained only boron for the purpose of lowering the melting point. Strong void-free joints were obtained by using an interlayer which contained both boron and silicon, and which slightly melted the base-metal surfaces at the joint. By using such an interlayer, it was also possible to bond A286 to nickel-based Inconel-713C[215].

AISI304

This stainless steel was bonded by using a pure copper or aluminium foil as an interlayer and heating under vacuum. The hardness of the bond made by using a copper interlayer was higher than that of one made using an aluminium interlayer. Poor mechanical properties of the bonds were associated with the formation of intermetallics within the bond region. Upon holding the parent alloy for longer periods, complete isothermal solidification tended to occur. Diffusion of the main elements from the interlayer and into the base metal at the bonding temperature was the main factor controlling microstructural evolution of the joint interface. The choice of a suitable bonding temperature to give the highest concentration of melting-point depressants depended upon the isothermal solidification time. The presence of eutectoid-γFe + eutectic Cu+Cr and Fe-Al intermetallic was detected at the interface of joints bonded using copper and aluminium interlayers, respectively[216]. The copper diffusivity in the grain boundaries of the steel was about 56×10^5 times higher than the lattice diffusivity at the joint interface, showing copper - as a melting point depressant – was able to produce grain-boundary grooves and facilitate diffusion of the copper atoms. The highest microhardness was found at the diffusion zone, and decreased gradually with increasing distance from the joint. When Tafel tests were performed using 3.5%NaCl solution, joints involving copper developed crevice corrosion due to galvanic couples[217]. Pitting also occurred due to intergranular stress corrosion cracking at the copper surface. The steel has also been bonded by using a nickel-based BNi-9 interlayer. It was found that silicon nano-particles acted as a melting-point depressant during bonding, with the formation of a silicon transient liquid phase leading to dissolution of elements in the interlayer and to a uniform distribution in the bonding area. Silicon nano-particles led to the appearance of smaller eutectic structures in the molten zone[218]. The microhardness was lower when silicon nano-particles were used in the bonding process. The austenitic stainless steel was bonded by using a 40μm-thick cobalt-based interlayer. Isothermal solidification was complete within 0.5h at 1180C. With increasing homogenisation time at 1000C, there was a more uniform distribution of the alloying elements and of the hardness across the joint region[219]. The average shear strength of homogenised samples was equal to about 72% of that of the base metal. Powder particles which are coated with a small amount of melting-point depressant exhibit a different sintering behavior when compared with that of a non-coated powder mixture having the same composition. Interlayers which consisted of such coated powder particles have long been used for transient liquid-phase bonding. The nature of the coating material and the deposit thickness were important factors governing shrinkage. The amount of melting-point depressant was chosen so that the volume fraction of melt was very small but was present at all points of contact in order to improve densification. When Ni-20Cr and 304L stainless steel powders, coated with Ni-10P, were used to join 304 stainless steel fully dense joints which possessed mechanical properties comparable to those of the base metal were obtained using Ni-20Cr powder. Joints made using 304L stainless steel powder interlayer possessed inferior properties due to the presence of residual porosity in the joint [220].

Fe-Cr-W

Austenitized and tempered martensitic steel with the composition, Fe–0.11C–8.5Cr–1.5W–0.29V–0.53Mn–0.10Ta–0.025wt%N, was bonded by using a 50μm-thick amorphous interlayer of Ni–Si–B–Fe–C foil and heating at up to 1060C. As the temperature reached the start temperature for austenite transformation, the boron melting-point depressant in the foil could diffuse along the austenite boundaries to form compounds and fine-grained zones adjacent to the interlayer. Due to melting of the interlayer and to dissolution of the substrate, the final width of the interlayer increased [221]. Small voids were observed at the interlayer close to the joint interface and were attributed to the difference in iron and nickel diffusion rates. The bonded steel fractured at the interlayer.

TP304H

This stainless steel has been bonded by using nickel-based, iron-nickel based or iron-based interlayers. A nickel based interlayer produced a nickel solid solution with continuous growth of the grains across the joint. The bond failed at the joint, but the tensile and bending strengths were equal to those of the base metal. The microstructure of a joint produced using a iron-nickel based interlayer was different to that of the base metal, and diffusion between the interlayer and the base metal was poor. The bond failed at the joint, and the tensile and bending strengths were lower than those of the base metal [222]. An iron-based interlayer produced an homogenous joint having a microstructure and composition which were similar to those of the base metal. In tensile tests, the bond here failed in the base metal instead of at the joint.

UNSS31803

This nitrogen-containing duplex stainless steel has been bonded by using a Ni-7Cr-3Fe-4.5Si-3.2wt%B amorphous interlayer and by heating at 1283 to 1353K for up to 1000s under vacuum. The austenite volume fraction decreased with increasing temperature and holding time. After prolonged heating, depleted areas of austenite were observed in the base metal adjacent to joints. There was a linear correlation between the width of the remaining liquid and the square root of the holding time at a given temperature. The secondary phases which were found in the joint area were mainly (Cr,Mo) borides. When the bonding time was longer than 1000s, boron nitride formed at the center and interface of the joint area, but its amount was decreased in comparison with shorter bonding times. The tensile strength increased with the holding time, and the bonding efficiency was about 94% for 1000s at 1353K[223]. The tensile strength was controlled by brittle eutectic and borides at short holding times, and by boron nitride - formed at the joint interface - following complete isothermal solidification.

FeCr-Ni

When a duplex stainless steel was bonded using a nickel-based amorphous interlayer, the dissolution width increased with increasing holding-time and attained a relatively stable value. The dissolution width increased as the bonding temperature was increased. The effect of the joint gap upon the dissolution-width was small. On the basis of the Nernst-Brunner equation, the dissolution activation energy was calculated to be about 282.4kJ/mol. This was less than the corresponding values for nickel-based alloys, and was attributed to the greater number of grain boundaries in the duplex stainless steel[224]. Transient liquid-phase bonding of duplex stainless steel has also been performed by using a Ni-Cr-B alloy insert. Before the completion of isothermal solidification, the bond region was of the form, γ-Fe+δ-Fe|γ-Fe+δ-Fe+BN|γ-Ni(Fe)+BN|γ-Ni+Cr-rich borides|γ-Ni+Ni$_3$B+Cr-rich borides (CrB, CrB$_2$, Cr$_2$B$_3$, Cr$_3$B$_4$, Cr$_5$B$_3$ and CrB$_4$), in going from the base metal side to the bond|interlayer side. Complete isothermal solidification occurred within 1h at 1090C. The γ-Ni solid solution phase was alone present in the bonded interlayer, and BN precipitates were not removed after isothermal solidification. The presence of peak micro-indentation hardness levels in the bond region was attributed to the formation of secondary-phase precipitates[225]. A duplex stainless steel was bonded by using MBF-30 (Ni-4.5Si-3.2wt%B) or MBF-50 (Ni-7.5Si-1.4B-18.5wt%Cr) interlayers. When using MBF-30, the microstructure of the athermal solidified zone depended upon boron diffusion at 1060C. A supersaturated γ-Ni phase and Ni$_3$B were observed in this zone, while BN appeared in the bonding-affected zone. When using MBF-50, the effect of base-metal elements as well as silicon from the interlayer upon the bond microstructure was more marked at 1175C. Here, BN and $(Cr,Ni)_3Si$ phases were present in the bond center-line[226]. The formation of BN precipitates in the bonding-affected zone was suppressed[227]. An appreciable deviation of the isothermal solidification rate from the predictions of normal bonding diffusion-models was found for joints prepared at 1175C using MBF-50.

MA956

It has long been known that an Fe-B-Si foil interlayer is suitable for bonding this iron-based iron-oxide dispersion-strengthened alloy. Such bonding produces joints which are free from intermetallic precipitates and are identical in composition to the parent metal. The use of nickel-based foil led to an austenitic bond region which was stabilized by a high nickel concentration. Retention of melting-point depressants such as silicon, at the centre of joints, resulted in the formation of silicide-boride precipitates at the bond centre and at the braze|parent-metal interface. High-temperature heat-treatment then failed to remove γ-Fe phase and precipitates, and their presence degraded the mechanical properties. The formation of intermetallic precipitates at the braze centre was attributed to the initial high concentration of chromium in the nickel-based brazing foil. The strength of joints made using iron-based foil was greater than of those made using nickel-based interlayers[228]. When iron-based foil was used, the bond strength at room temperature and at 700C was close to the parent-metal strength. At room temperature, failure occurred away from the bond interface. At high temperatures, the joint failed along the bond interface and this was attributed to melt-back. In order to decrease the completion time during bonding, an iron-based interlayer metal consisting of MA956 plus 7wt%Si and lwt%B was developed. Joints which were free from microvoids and bond-line intermetallics were obtained by bonding at 1563K under a pressure of $7.0MN/m^2$. The bond-line region in the base metal had a bamboo-like microstructure. During tensile testing at 923K, joints which were bonded at 1563K for 2.16ks fractured in the base-metal zone and the mechanical properties of the joint region and the base metal were similar[229]. The creep rupture properties of the joint region were close to those of the base metal in the transverse direction. The initial melting and homogenisation of the liquid interlayer can result in parent metal dissolution and to the loss of dispersion-strengthening phases from the joint region. A theoretical study showed that the extent of dissolution was determined by the thickness of the interlayer used, and it was suggested that thinner interlayers in the form of sputter coatings would greatly reduce dissolution[230]. Experimental results showed that 2μm-thick sputter coatings based upon the Fe-B-Si ternary system could be deposited by means of radio-frequency sputtering. The resultant bonds tend to be free from intermetallic precipitates and pores and exhibit no agglomeration of the dispersion strengthening phase. This is attributed to the lower volume of boron and silicon deposited in the coatings. The mechanically alloyed ferritic superalloys, MA956, MA957 and PM2000, were bonded by using an amorphous Fe-B-Si foil or a sputtered Fe-B-Si deposit. A boron-induced secondary recrystallized zone was found, at the bond line in MA957, which acted as a barrier to further grain growth across the bond line during subsequent annealing. Differential scanning calorimetry showed that this recrystallization was triggered at some 200C below the usual recrystallization temperature during heat treatment, and occurred only when the metal filler melted so that there was an easy flux of boron into the base metal. The boron-induced recrystallization was of the same nature as that produced by heat treatment, but led to a stronger <110> fibre texture. Grain growth across the bond line occurred in the bonds produced in MA956. Similar heat treatment of PM2000 produced a simultaneous but independent secondary recrystallization at the joint and in the bulk[231]. This difference in alloy behaviour was attributed to differences in their thermomechanical processing. Physical vapor deposited boron thin films have continued to be used as interlayers for bonding the oxide-dispersion strengthened alloys[232]. Better microstructural continuity occurred when the substrates were in the unrecrystallized fine-grain condition, and with the faying surface aligned in the working direction; as compared to when the substrates were aligned in the direction normal to the working direction[233]. The related iron-based oxide-dispersion strengthened superalloy, MGH956, could be joined by using a suitable interlayer. Joints which were free from residue were obtained by heating at 1240C for 8h. Increasing the holding time homogenized the joint microstructure and made it the same as that of the base metal[234]. Secondary phases were distributed on the joint surface. The high-temperature tensile properties were improved but variable. Porosity could not be eliminated by increasing the holding time, and its presence was related to the preparation details and to the cleanliness of the bonding surface[235].

Nickel-Based

CMSX

The monocrystalline superalloy was joined, using MBF-80 or F-24 interlayers, by heating at 1373 to 1548K for 1500 to 1800s in vacuum[236]. The [001] orientation of each specimen was aligned perpendicular to the bond interface, and the dissolution-width of the base metal at the bonding temperature increased with increasing temperature or time. The dissolution of the base metal in the molten interlayer could be described in terms of Nernst-Brunner theory. The eutectic width decreased linearly as the square root of the holding time during isothermal solidification. The quantity of micro-constituents in the bonded layer following isothermal solidification decreased with bonding temperature, and disappeared upon heating at 1523K for 0.5h when MBF-80 filler was used[237]. The tensile strength of the joint at high temperatures was equal to, or greater than, that of the base metal between 923 and 1173K. The creep rupture strengths and rupture lives of joints were almost identical to those of the base metal. The low-cycle fatigue properties of joints were also of the same magnitude as those of the base metal[238]. The tensile fracture surfaces of the joints were all similar in morphology, and failure always occurred in the base metal. Homogenization of the joints of CMSX-2 and CMSX-4 was investigated after using MBF-80 and F-24 interlayers and heating at 1523 to 1548K in vacuum for 1800 to 2400s. The (100) orientation of the bonded specimen was again aligned perpendicularly to the joint interface. Homogenization of the bond region occurred quickly when the bonding temperature was increased. The microstructures and hardness distributions of the bond interlayer, following homogenization at 1523K for 0.5h, were the same as those of the base metal. Analysis of the bond region, following solution treatment at 1589K for 2h - without any homogenization treatments – showed that joint homogenization occurred during the solution treatment[239]. The homogenization treatment could be replaced by sequential post-bonding heat treatment. The effects of the base-metal grain size upon dissolution and isothermal solidification of monocrystalline, coarse-grained and fine-grained CMSX-2 were investigated after using an MBF-80 interlayer and heating at 1373 to 1523K in vacuum. The dissolution of the base metal conformed to Nernst-Brunner theory regardless of the base-metal grain size. The saturation time for dissolution, and the saturated dissolution width of the base metal, decreased in the order: monocrystal, coarse-grained, fine-grained. The eutectic width decreased linearly as the square root of the holding-time during isothermal solidification for monocrystalline, coarse-grained and fine-grained base metals. The time required to complete isothermal solidification decreased in the order: monocrystalline, coarse-grained, fine-grained[240]. The differences in behaviour were attributed to the effect of the base-metal grain boundaries upon boron diffusivity in CMSX-2. Joints in monocrystalline CMSX-2 have been investigated three-dimensionally. The (200) pole-figure and stereographic triangles indicated that all of the analyzed points had almost the same orientation across the joint interface. The misorientation was negligible for as-bonded and post-bonding heat-treated conditions. The atoms were arranged continuously across the bond interface and were quite coherent at the bond interface[241]. It was concluded that epitaxial growth of the solid phase started from the base-metal substrate during isothermal solidification. When bonding of CMSX-2 and CMSX-4 was performed using MBF-80 and F-24 interlayers and heating at 1373 to 1548K for up to 19600s in vacuum, the dissolution width at the bonding temperature increased as the temperature and time were increased[242]. The eutectic width again decreased linearly as the square root

Materials Research Forum LLC
doi: http://dx.doi.org/10.21741/ 9781644900055

of the holding time during isothermal solidification. The bond interlayer became monocrystalline during the bonding process and matched the orientation of the bonded substrates[243]. The levels of high-temperature tensile strength and of creep-rupture strength of the joints were identical to those of the base metals. Monocrystalline CMSX-2 was again bonded by using an MBF-80 interlayer, and the resultant bond misfit was characterized in terms of the <100> twist-angle at the joint interface. Post-bonding heat treatment was carried out in argon, or under vacuum, before creep-testing[244]. The joint creep-rupture properties were comparable to those of the base metal, for twist angles of up to 3°, but worsened sharply when that angle exceeded 5°. The boundary energy and oxygen content at the bond interface increased with increasing twist angle[245]. These creep rupture properties could be improved by lowering the oxygen partial pressure during post-bonding heat treatment, and this was attributed to a reduction in grain-boundary oxidization of the bonded interface. The elongation and reduction-in-area values were comparable to those of base metal, while the creep rupture at high temperature was worsened. The low-cycle fatigue properties of the joints were also the same as those of the base metal[246]. The fracture surfaces following creep rupture tests could be divided into three regions, and the elongation and reduction-in-area values decreased sharply with increasing areal fraction of the interfacial fracture surface. When monocrystalline CMSX-4 was bonded by using MBF-80 as an interlayer and heated at 1200C for 3 to 6h, followed by hot isostatic pressing and heat treatment, samples which were bonded at 1200C for 3h and subjected to a three-stage heat-treatment exhibited superior properties[247]. When as-cast nickel-based rhenium-containing CMSX-4 was bonded by using a hafnium-containing nickel-based interlayer, bonding was complete after heating at 1290C in vacuum for 24h[248]. No diffusion-affected zone was found.

As well as for joining CMSX-4, the Ni-Cr-B interlayer alloys could also be used for repair. The microstructure of the repaired region could be divided into an eutectic zone, a normally solidified zone, an isothermally solidified zone, a precipitate zone in the base metal and the substrate[249]. The isothermally solidified zone and the precipitate zone in the base metal exhibited exactly the same crystal orientation as that of the base metal. The dissolution width of the base metal, and the isothermal solidification width, increased with increasing repair temperature and holding time. Built-up repairs of up to 38μm in width could be performed by heating at 1448K for 49h, using the Ni-Cr-B filler. Further study of the technique, using Ni-Cr-B and Ni-Cr-Si alloys as fillers and applied to monocrystalline CMSX-4, revealed that good results were obtained by using Ni-4Cr-10Si filler and heating at 1473K for 25h[250]. The creep rupture strength of parts repaired with Ni-4Cr-10Si filler attained some 67% of that of the base metal.

DD3

This nickel-based monocrystalline superalloy has been bonded by using D1P powder or D1F amorphous foil as an interlayer. There was a clear microstructural inhomogeneity of the joints which were made using D1P, and long-term diffusion treatment at high temperatures was required in order to remove the inhomogeneity. When bonding using an amorphous foil interlayer, it was easier to obtain a joint microstructure which was identical to that of the base metal[251].

DD6

The monocrystalline superalloy was bonded by using an interlayer having a composition which was similar to that of DD6, plus a certain amount of boron addition as a melting-point depressant. It was difficult to obtain joints having completely homogeneous microstructures. After heating at 1290C for 12h, some 50% of the sample exhibited a $\gamma+\gamma$' microstructure which was almost identical to that of the base metal[252]. The remainder of the sample consisted of γ-solution and borides, plus other constituents. Upon increasing the bonding time to 24h, the inhomogeneous areas of the joint were reduced, and the properties were improved. The stress-rupture strength attained 90 to 100% and 70 to 80% of those of the base metal at 980 and 1100C, respectively[253]. Computer-aided alloy-design has been used to choose interlayers for the joining of $\gamma/\gamma'/\beta$-type high-aluminium nickel-based superalloys. The optimum chemical composition choice involved making a compromise between competing factors, and an index was defined which took account of melting points, hardness, brittle-phase formation tendency and void ratio in the joint. This led to the chemical composition, Ni-30Cr-40B-0.5wt%Ce (christened HY-20), being designated as the optimum one for the interlayer. Joints made using this interlayer exhibited good microstructures, with no formation of brittle phases or voids[254]. The creep rupture properties of joints made using this interlayer alloy were also greatly improved, as compared with joints made using amorphous MBF-80 (Ni-15.5Cr-3.7B), and were quite comparable to those of the base metal.

DD22

Optimum bonding conditions produced stress-rupture properties at 980C which could exceed 90% of those of the base metal. With further development, the joint comprised a bonding zone and a parent-material zone; with the diffusion zone being hardly visible. The bonding zone in turn contained an isothermal solidification zone and a residual liquid zone[255]. The joints required post-welding heat-treatment, and the mechanical properties of the joint could be improved by limiting the formation of grain boundaries in the diffusion bonded joint.

Transient Liquid Phase Bonding   Materials Research Forum LLC
Materials Research Foundations **43** (2019)   doi: http://dx.doi.org/10.21741/ 9781644900055

DZ22

The directionally solidified nickel-based superalloy was bonded by using Z2P powder or Z2F amorphous foil interlayers. Joints made using the Z2F amorphous foil were clearly more homogenous than were joints made using the Z2P powder. Joints made using the amorphous foil also exhibited better stress-rupture properties; ones which attained 90% of those of the base metal[256].

FGH97

The nickel-based superalloy was bonded by using a BNi82CrSiB interlayer and heating at 1150C for various times. As the time was increased, the width of the bond region broadened, and the amount of γ solid solution in the isothermally solidified zone gradually increased. On the other hand, the amounts of dendritic γ/Ni_2B eutectic, γ/γ' eutectic, Ni_3Si, $M_{23}B_6$ and Cr_2B phases in the non-isothermally solidified zone gradually decreased and finally vanished. A microstructure which consisted of compacted γ solid solution finally formed. Acicular, granular and vermiculate M_3B_2 phases, enriched in chromium, tungsten, cobalt and molybdenum, precipitated in the diffusion zone and the latter's width increased with increasing holding time[257]. Interdiffusion between the molten interlayer and the base metal occurred during bonding, resulting in local melting of the base metal and to extensive precipitation of borides in the diffusion-affected zone.

GTD-111

This nickel-based superalloy was bonded by heating to between 1373 and 1423K in vacuum, at a rate of 0.1, 1.0 or 10K/s. When the heating-rate was low and the bonding temperature high, the time required for the completion of dissolution after reaching the bonding temperature decreased. When the heating-rate was very low, solidification proceeded even before reaching the bonding temperature, and the time required for the completion of isothermal solidification was shortened[258]. When the total time required for the completion of solidification, right from the beginning of heating was considered, heating at 0.1K/s had almost the same effect as heating at 10K/s. Directionally solidified GTD-111 was later bonded at 1403K and the liquid interlayer was removed by an isothermal solidification which was governed by the diffusion of boron and silicon into the base metal. Solids in the bonded interlayer grew inwards, simultaneously and epitaxially, from the interlayer and contacting base metal. The numbers of grain boundaries which formed at the bonded interlayer corresponded to those of the base metal. Melting at the grain and dendrite boundaries occurred at 1433K. At a bonding temperature of 1453K, higher than the melting point of the grain boundary, the liquid of the interlayer was contiguous with molten grain boundaries. This continuity extended as far as the grain boundary; some 1.5mm from the interface. This liquid was a mixture of the interlayer metal and the phase which existed at the grain boundary[259]. During extended holding, the liquid phases gradually disappeared and those originally in a continuous band formed discrete islands. The liquid phases did not disappear during holding at 1453K for 2h. Extended isothermal solidification at a bonding temperature which was higher than the melting-point of the grain boundary was governed by titanium diffusion. The superalloy has also been bonded by using an MBF30 (Ni-Si-B) amorphous interlayer and heating at 1100C under vacuum. Before the completion of isothermal solidification, the bond region comprised 4 distinct zones: a center-line eutectic structure due to athermal solidification, a solid-solution phase due to isothermal solidification, diffusion-induced boride precipitation … and the base metal[260]. Complete isothermal solidification prevented the formation of center-line eutectic, and occurred within 1.25h at 1100C. The homogenization of isothermally solidified bonds at 1150C for 4h led to a reduction in the number of secondary precipitates in the diffusion-affected zone and to the formation of significant numbers of γ' precipitates in the bond region. Estimates have been made of the time required to obtain an intermetallic-free joint center-line during the bonding of the cast nickel-based alloy by using an amorphous Ni-4.5Si-3.2wt%B interlayer[261]. By taking account of the solidification behaviour of the residual liquid, the isothermal solidification time could be predicted by solving a time-dependent diffusion equation based upon Fick's second law. When using an MBF-50 interlayer and a gap of less than 100μm, bonding and homogenization are lengthy. As already noted, isothermal solidification is controlled by the diffusion of boron and silicon. The amounts of those elements in MBF-50 are however high. This explained the lengthy bonding time. The MBF-50 also does not contain aluminium or titanium, which are γ'-phase formers[262]. This explained the long homogenization time. Modified interlayers were developed which contained aluminium and titanium as γ'-phase formers and had reduced boron. When the new composition was used, the bonding time was decreased by about 1/10 and no homogenization treatment was required. Even without the latter, the volume fraction of γ'-phase in the joint interlayer was equal to that in equivalent homogenized material.

As mentioned elsewhere, transient liquid phase bonding is now an essential technique for the repair of gas-turbine components. Two typical γ'-precipitate-strengthened superalloys which might well require repair are directionally solidified GTD-111 and wrought Udimet520. Bonding has been carried out using an amorphous interlayer under various conditions and the results clearly showed that joints in these superalloys exhibited different bonding behaviors, depending upon the temperature used, and that the presence of dendritic structures and the melting point of the base metal had important effects upon their bonding behaviors[263].

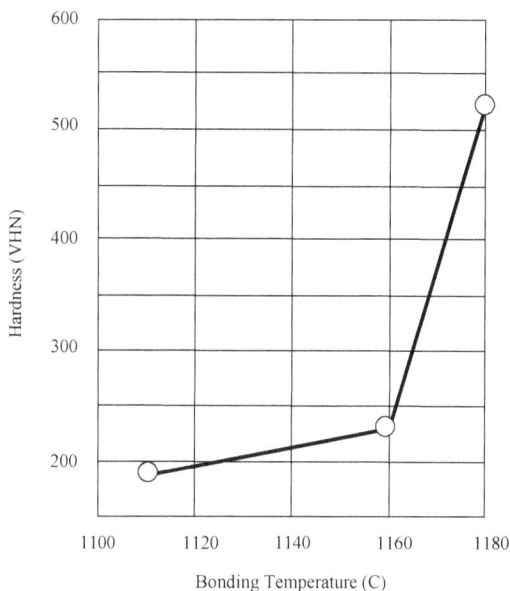

Figure 17. Joint hardness of Haynes-282 as a function of bonding temperature

Haynes-230

The nickel-based alloy, Haynes 230, was bonded by using Ni-6wt%P or Ni-12wt%P as an interlayer and heating at 1150C under 12.7MPa. The microstructures had three distinct regions: a joint center-line which contained pores, an isothermally solidified zone free from carbide precipitates and the base alloy. There were no deleterious phases at the joint or in the isothermal solidification zone. The yield strength attained values of up to 86% of the original alloy sheet when tested at 750C. The elongation to fracture was negligible, but the fracture surfaces exhibited ductile cup-and-cone failure passing through the boundary between the isothermal solidification zone and the joint. Plastic strain was limited to the region through which fracture occurred[264].

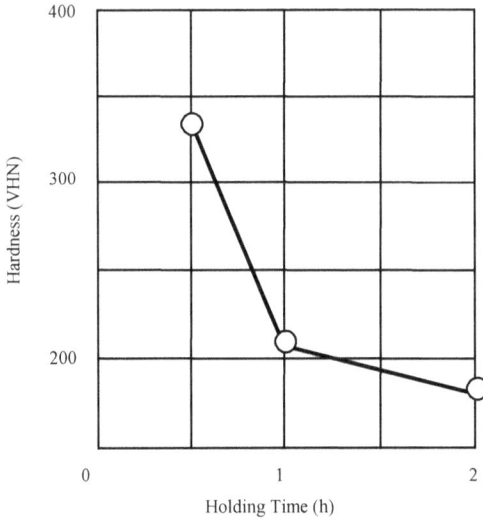

Figure 18. Joint hardness of Haynes-282 as a function of holding time at 1100C

Haynes-282

During the joining of Haynes 282, an insufficient holding-time for the complete isothermal solidification of a molten interlayer led to the formation of eutectic-type micro-constituents along the joint center-line. An increase in bonding temperature to above a certain temperature unexpectedly led to a decrease in the isothermal solidification rate and to the formation of the centreline eutectic whose existence was avoided by using the same holding-time at a lower temperature[265]. Finite-element numerical simulation showed that this anomalous behaviour could be attributed to a deviation from the parabolic relationship between solid|liquid interface migration and holding time which was in turn caused by a reduction, to below a critical value, of the solute concentration gradient in the solid substrate. A comparison of these results with those for nickel-based superalloys such as EN738 showed that the undesirable reduction in isothermal solidification rate was much less marked in Haynes282. This was attributed to the more extensive formation of interfacial molybdenum-rich borides in the new alloy. This could relieve boron saturation in the matrix and reduce the degree of the above deviation[266]. The joint hardness increased sharply with bonding temperature (figure 17) but decreased with holding time (figure 18).

IC6

This Ni_3Al-based directionally-solidified superalloy has been bonded by using a Ni-Mo-Cr-B powder interlayer. The joint microstructure consisted of a mixture of γ solid solution, or of a γ+γ' structure plus borides[267]. With increasing bonding time, the number of borides in the joint and adjacent zones gradually decreased and the joint stress-rupture properties improved. The latter were comparable to those of the superalloy. Monocrystalline Ni_3Al was bonded by using a KNi3 interlayer and heating at 1240C for 12h. Most of the joint was the same as the matrix and comprised γ and γ' phases. Other parts of the joint consisted of a γ+γ' eutectic phase plus small blocky borides. The γ'-phase of the base metal changed in shape from largely cuboidal to irregular[268]. The stress rupture strength of the joint at 1000C attained 90% of the value for the base metal. During stress rupture, the γ'-phase in the bond area was coarse and irregular, with some angles being aligned with the direction of stress rupture.

IC10

Joints of the bonded monocrystalline superalloy comprised a bond zone and a base-material zone. The former consisted in turn of an isothermal solidification zone and a rapid solidification zone. The latter could be avoided by increasing the bonding time. The size of the γ'-phase particles in the base material reached 0.9μm when the joining time was increased from 2 to 8h. The mechanical properties of the joints could be improved by limiting the formation of grain boundaries in the diffusion-bonded joint and by post-joining solution-treatment[269]. The average tensile strength of the joints was 507MPa at 1000C, and the creep rupture life attained 120h at 1000C under a 144MPa stress.

Inconel-617

This superalloy was first bonded by using Ni-4.5Si-3wt%B, where boron was the melting-point depressant, as the interlayer and heating at 1065C. Two different thickness of interlayer and various holding times were used. The presence of MoB, Mo_2B, $M_{23}C_6$, TiC, $M_{23}(B,C)_6$ and Ni_3B precipitates was detected in the diffusion layer, while Ni_3B, Ni_3Si and Ni_5Si_2 precipitates were found in the interlayer, at the interface between the latter and the base metal[270]. Another method was to use a 25.4 or 38.1μm-thick Ni-11wt%P interlayer, where phosphorus was the melting-point depressant, and heat at 1065 or 1150C for 300s to 24h. A uniform microhardness and microstructure across the joint was obtained using a 25.4μm-thick interlayer and heating at 1150C for 24h[271]. The alloy

Materials Research Forum LLC
doi: http://dx.doi.org/10.21741/9781644900055

has also been bonded by using a 50 or 100μm-thick BNi-1 interlayer and heating at 1135C for 5h in vacuum, argon or air. After 5h, isothermal solidification was complete for any thickness or atmosphere, but joints which were made under vacuum had the highest shear strength, failure energy and ductility. Joints made in argon, using a 100μm-thick interlayer, had a good strength but those made in air were weak, due to the presence of cavities and undissolved carbides. On the other hand, the presence of such carbides in the diffusion-affected zone resulted in an increased hardness of the base metal near to the interface[272]. The carbides dissolved, and the hardness of the diffusion-affected zone was reduced, for vacuum-bonding conditions as compared with the results for bonding in argon or air. The best result, equalling some 95.8% of that of the base metal, was obtained by bonding under vacuum using a 50μm-thick interlayer.

Inconel-625

Incidental to the joining of this superalloy to itself, a cyclic method has been developed for measuring the rate of isothermal solidification of joint-centerline eutectic in a nickel-based couple by using a single sample. It involves determining the solidification enthalpy of the remaining eutectic liquid in a half-bonded joint when cooled from the bonding temperature. Rapid diffusional solidification occurs during the initial melting to the extent that about half of the eutectic liquid has already diffusionally solidified before the maximum bonding temperature is reached[273]. The initial diffusional solidification thus greatly reduces the total time required for solidification at the bonding temperature.

Inconel-718

Fine-grained Inconel 718SPF superalloy sheets were first bonded by using a Ni-P or Ni-Cr-P amorphous interlayer. A joint having a uniform chemical composition could be obtained by using a Ni-P interlayer and heating at 1100C for 8h. When a Ni-Cr-P interlayer was used under the same conditions, the concentrations of nickel, iron and niobium in the bond region and in the base metal exhibited differences of more than 2wt%. This implied that a longer bonding time was required to homogenize the chemical composition of joints with a Ni-Cr-P interlayer. Joints with a Ni-P interlayer had a higher strength than did those with a Ni-Cr-P interlayer[274]. In an analogous Bayesian expert-system study of the bonding of Inconel-718 the composition, Ni-3.0Mo-6.4Cr-4.7Nb-2.0Si-1.0wt%B, was determined to be the optimum one for an interlayer. The processing parameters were here optimised with regard to joint strength, reduction-of-area and toughness, and the best treatment was 3300s at 1459K[275]. No brittle phases were then formed in the bond layer, and the joint strength and toughness were better than 80% of those of the base-metal. When wrought IN718 was bonded, standard solution treatment at

1000 or 1050C for 1h could ensure the completion of isothermal solidification of the IN718|Ni-Cr-Fe-Si-B|IN718 system, leading to an entirely intermetallic-free joint center-line. Two distinct microstructural zones formed in the bond zone: an isothermal solidification zone, consisting of a single-phase solid solution, and a diffusion-affected zone which consisted of copious diffusion-induced boride precipitates. The aging responses of the isothermal solidification zone, diffusion-affected zone and base metal were different, due to the existence of a compositional gradient across the joint region[276]. The hardness of the isothermal solidification zone was the factor governing the shear strength of bonded IN718. When bonded by using an amorphous Ni-15.2Cr-4%B interlayer, and following partial isothermal solidification, the residual liquid in the joint center-line transformed, during cooling, into a non-equilibrium eutectic comprising nickel-rich borides, chromium-rich borides and an eutectic γ-solid solution phase[277]. Complete isothermal solidification, leading to an intermetallic-free joint centreline, occurred after heating for 40min at 1100C. Increasing the bonding time improved the joint shear strength due to a decrease in the width of the eutectic-type micro-constituents at the joint center-line[278]. Similar bonding was performed by using an amorphous Ni-7Cr-4.5Si-3Fe-3.2wt%B interlayer and heating at 1273 to 1373K. By taking account of the solidification behaviour of the residual liquid, the time required for isothermal solidification could be predicted by solving a time-dependent diffusion equation based upon Fick's second law. The aging behaviour of the joint was affected by the low Nb+Al+Ti content of the isothermal solidification zone and the formation of Nb-Cr-Mo-based boride precipitates in the diffusion-affected zone[279]. Post-bonding heat treatment could avoid diffusion-induced boride precipitation in the diffusion-affected zone and increase the Nb+Al+Ti content of the isothermal solidification zone. Progress of the essential diffusion-induced isothermal solidification stage of transient-liquid-phase bonding has been modelled by using the Larson-Miller parameter. The solidification of the liquid phase during bonding of wrought IN718 using various bonding times and temperatures was analyzed. Based upon that parameter, the bonding parameters could be defined by: $T_B[70 + \ln(t_B)]$ for a given bonding temperature, T_B, and time, t_B. There was a direct linear relationship between the Larson-Miller parameter and the size of the isothermal solidification size[280]. A study of microstructural changes during the post-bonding heat treatment of wrought IN718 for 12h at 1150C (lower than the solvus temperature of borides) revealed that there was fragmentation, break-up and dissolution of the Cr-Mo-Nb rich borides[281]. This dissolution of the borides at a temperature lower than their initial solvus temperature was attributed to a modification of the boride composition during diffusion-induced homogenization. Increasing the post-bonding heat treatment time resulted in enhanced solid-solution strengthening of the joint center-line and a greater aging response, due to alloying-element homogenization. Most recently, the superalloy has been bonded at 1050C by using a Ni-Cr-Fe-Si-B-based mechanically-alloyed equiaxed 12μm powder interlayer. The formation of metastable eutectic lowered the melting-point of the interlayer to 1025C. Homogeneous joints were produced by using this interlayer material. Three different zones were observed at the bond: isothermally solidified, diffusion-affected and unaffected base metal[282]. In the diffusion-affected zone, boron was present, at the grain boundaries of the nickel γ matrix, in the form of bulky borides. The diffusion of boron from the interlayer material and into the base material governed the isothermal solidification and bond formation.

Inconel-738

The nickel-based superalloy, IN-738LC, was originally joined by using AMS 4777 alloy as an interlayer and heating at 1050C under vacuum for various times. Continuous center-line eutectic phases, which could be characterized as being nickel-rich boride, chromium-rich boride and nickel-rich silicide, were found in joints with incomplete isothermal solidification. In addition to the eutectic products, the precipitation of boron-rich particles was also observed in the diffusion zone[283]. When the time was increased to 1.25h, the eutectic zone was completely eliminated and the joint isothermally solidified. Homogenization (1120C, 5h) of isothermally solidified joints led to the elimination of intermetallic phases which had formed in the diffusion-affected zone and to the formation of appreciable numbers of γ′ precipitates in the joint region. Plates of IN738 were later bonded by using commercial brazing alloys. It was found that a slight degree of wetting was associated with the rapid onset of isothermal solidification, whereas a large degree of wetting was associated with a relatively low rate of isothermal solidification[284]. Borides which exhibited a so-called blocky morphology were present in all of the joints, whereas a Ni-Ni₃B eutectic phase was observed when using Nicrobraz 150, while the others led to the appearance of a coarse γ′-phase with no Ni₃B matrix phase. Cast Inconel-738LC was bonded by using a Nicrobraz-150 (Ni-Cr-B) interlayer. Isothermal solidification occurred in two separate regimes, depending upon the temperature used. The rate of isothermal solidification was faster in the first regime than it was in the second regime. This led to a deviation, from then current theory, with regard to the isothermal solidification completion-time[285]. Such a deviation was suggested to constitute an alternative explanation for the anomalous prolongation, with increasing bonding temperature, of the holding time required to produce an eutectic-free joint[286]. The change in solidification rate was attributed to an appreciable enrichment, of the liquid interlayer, in base-alloy elements and its continual change during isothermal solidification. The process was also affected by the nature of the phases which formed in the center-line eutectic following incomplete isothermal solidification[287]. Those phases could include chromium-rich M_5B_3, nickel-rich $M_{23}B_6$ and nickel-based γ solid solution. An appreciable volume fraction of complex face-centered cubic Cr-Mo-W rich carboborides was also observed in the joint|base interface region. It was suggested that solid-state diffusion of boron, before the completion of equilibration, led to the formation of carboboride phases. Modelling of the bonding of IN-738 using Ni-Cr-B filler once assumed that boride precipitation did not occur during the process. Extensive intragranular and intergranular precipitation of chromium- and boron-rich particles was however observed to occur, within the base alloy, next to the interlayer|base-metal interface. Diffusion models were successful in predicting quite accurately the time required to complete the isothermal solidification of the liquid interlayer[288]. This knowledge was essential in order to prevent the formation of deleterious centreline eutectic. Such diffusional analysis also showed that boron-rich particles formed due to extensive solute diffusion into the base metal before equilibration of the liquid-solid phases occurred. When Inconel-738LC was bonded by using an Amdry DF-3 interlayer, the formation of eutectic micro-constituents within the joint regions was markedly affected by the temperature and time used. A deviation from then-current liquid-phase-bonding diffusion models was observed in samples which had been heated at above 1175C. The rate of isothermal solidification was greatly reduced at this temperature and also resulted in the formation of center-line eutectic micro-constituents which were different to those found at lower bonding temperatures[289]. A possible explanation for this was a considerable enrichment, in titanium from the base alloy, of the molten interlayer due to a lower partition coefficient in nickel-based alloys. The bonding of IN738LC was optimised by using a Bayesian expert system, leading to an index, of bonding performance, which involved the melting-point of the interlayer and the hardness of brittle phases in the bond layer. The composition, Ni-3.0Cr-8.1Si-1.0wt%B, was deduced to be the optimum one for the interlayer[290]. The process was optimised with regard to joint strength and toughness, and the best treatment was 5600s at 1433K. Brittle phases did not form in the bond layer when using this new interlayer, and the joint strength and toughness were better than the average base-metal properties. The redistribution of alloying elements in the bonding system, IN738LC|BNi-3|IN738LC, was studied in order to understand the microstructural evolution. During non-isothermal solidification in the bonding zone, enrichment of the residual liquid phase in the positively segregating elements led to the formation of intermetallics in the joint zone. The redistribution of alloying elements between the bonding zone and base alloy resulted In the formation of a γ' boundary layer, containing a high density of fine γ', around the bonding zone. The diffusion of γ'-promoting elements from the base alloy and into the bonding zone, controlled the growth of that layer[291]. Contrary to bonding at 1100 to

1150C, the isothermal solidification kinetics were notably reduced during bonding at 1200 to 1250C. These slow kinetics were not due only to the enrichment of the liquid phase in base alloying elements such as titanium and zirconium, but also to a reduction in the solid solubility limit of boron in the base alloy. Microstructural evolution in the bonding area was markedly affected by the bonding temperature, and a critical bonding temperature could be defined for the bonding of IN-738LC|BNi-3|IN-738LC samples. Unlike bonding below the critical temperature, bonding above it resulted in the formation of individual γ-γ' colonies in the base alloy adjacent to the bonding zone and caused a significant reduction in the isothermal solidification kinetics[292]. During bonding at temperatures lower than the critical one, boron-rich precipitates with blocky or acicular morphologies, formed in the base alloy adjacent to the bonding zone. Contrary to existing analytical models, which assumed a parabolic relationship between liquid|solid interface migration and holding time, deviations from that law were observed experimentally and in numerical simulations. The deviation was caused by a reduction in the solute concentration gradient to below a critical value and was suggested to be as an alternative explanation for the anomalous extension of the processing time (figure 19) required to produce an eutectic-free joint with increasing bonding temperature[293].

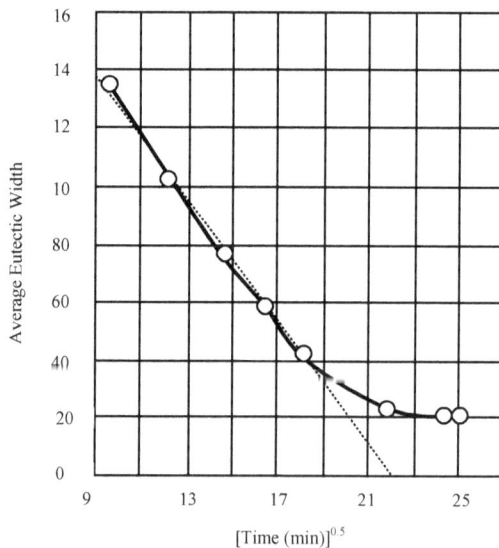

Figure 19. Eutectic width as a function of holding time of an Inconel-738 joint prepared at 1140C

Decreasing the interlayer gap, and using a melting-point depressant solute having a higher solubility in the base material could reduce the occurrence of this anomalous behavior. When IN-738LC was bonded by using an AMS4776 (Ni-B-Si-Cr-Fe) interlayer, containing boron and silicon as melting-point depressants, and heated at 1100C for various times, insufficient isothermal solidification of the molten interlayer led to the formation of continuous centreline eutectic phases. The width of the eutectic decreased with increasing holding time. Complete isothermal solidification, which prevented the formation of the centreline eutectic, occurred within 0.75h. In addition to centreline eutectic, the precipitation of boron-rich particles occurred in the diffusion-affected zone[294]. Homogenisation (1120C, 5h) of isothermally solidified joints resulted in the elimination of any intermetallic phases from the diffusion-affected zone, and in the formation of many γ' precipitates in the joint region. The rate of isothermal solidification decreased upon increasing the temperature up to 1150C. This low isothermal solidification rate was due not only to an enrichment of the liquid phase in base alloying elements such as titanium, but also to a reduction in the solid solubility limit of boron in the base metal[295]. Under conditions where isothermal solidification was not complete, the eutectic constituent having the greatest hardness in the bond region was the preferential failure source. Homogenized joints had the highest shear strength. Bonding of IN738LC superalloy has been carried out by using rapidly solidified MBF-15 35 to 140µm-thick nickel-based foil and heating at 1130 to 1170C for 300s to 2h. The solidification sequence in the joint region involved the formation of γ solid solution in the isothermally solidified zone, followed by a ternary eutectic of $\gamma+Ni_3B+CrB$ and finally a binary eutectic of $\gamma+Ni_3Si$ in the athermally solidified zone. Fine Ni_3Si particles also formed via solid-state transformation within the γ matrix in the vicinity of eutectic products. A deviation from parabolic of the isothermal solidification kinetics occurred upon increasing the bonding temperature to 1170C, resulting in the formation of eutectic at the joint center-line. Samples with complete isothermal solidification exhibited the highest shear strength. Hard eutectic constituents acted as preferential failure sites and led to a marked reduction in the joint shear strength in samples with incomplete isothermal solidification[296]. When Inconel-738LC was bonded by using an AMS4777 powder interlayer and a 40 or 80µm gap, and heated at 1120C for 2h, aging (845C, 24h) was also carried out before or after bonding. The complete isothermal solidification time for a 40µm gap was 0.75h. The relationship between the gap size and the holding time was not linear but, with increasing gap size, the eutectic phase width became greater. In both aging cases, there were no morphological changes in the diffusion-affected zone precipitates. There was however an enrichment of the base-metal elements, titanium and aluminium, in the isothermal solidification zone of heat-treated samples; especially when aging was performed before bonding[297]. This increased the hardness of the centreline, and an increased shear strength was found together with ductile fracture surfaces[298]. A maximum hardness of 759HVN and maximum shear strength of 565MPa were also found after aging before bonding, using a 40μm gap. These improved properties were attributed mainly to solid-solution strengthening. Electrochemical analysis of the corrosion performance of this superalloy showed that bonding using a Nicrobraz 150 interlayer led to a marked reduction in the corrosion resistance due to the formation of non-equilibrium solidification-reaction micro-constituents in the joint region. This degradation could be entirely eliminated by using a composite interlayer which had previously been discounted as being useful for joining monocrystalline superalloys[299]. The effectiveness of the new interalloy even became greater as the aggressiveness of the environment increased. The wettability of various polycrystalline and monocrystalline nickel-based superalloys by a composite interlayer has been determined via hot-stage optical microscopy of sessile-drop specimens and spreading tests[300].

K417G

The nickel-based superalloy was bonded by using a 2.2mm gap and liquid-phase infiltration. This led to a non-uniform microstructure and voids, Cr_5B_3 and $(Cr,W)B$ were present. The tensile properties, especially the ductility, of the joints were lower than those of the base metal. The fracture surfaces showed that Cr_5B_3 was the main cause of failure, leading to the decrease in ductility[301]. The boride acted as a crack initiation site and also accelerated the crack propagation during deformation.

K465

Polycrystalline nickel-based superalloy was bonded by using an amorphous Ni–Cr–Fe–B–Si interlayer. Boron-rich and silicon-rich phases were found at the center of the seam after bonding at 1210C for 0.5h, and isothermal solidification was complete after heating at that temperature for 4h. The relationship between the average width of the remnant eutectic zone, and the bonding time at 1210C, was non-linear[302]. The tensile strength of the joint at both room temperature and 900C was comparable to the equivalent data for K465.

MA758

This oxide-dispersion strengthened nickel alloy, in the fine-grained or recrystallized state, was first bonded using amorphous Ni-Cr-B-Si as an interlayer and heated at 1100C for various times. The final joint grain-size was independent of the parent-alloy grain structure and bonding time. When the alloy was bonded in the recrystallized condition and given a post-bonding heat treatment at 1360C, the joint grain-size increased and the parent-alloy microstructure was continuous across the joint. When it was bonded in the fine-grained condition and then recrystallized at 1360C, the grains at the joint appeared to increase in size with increasing bonding time[303]. In general, the joint grains were larger than those produced when the alloy was bonded in the recrystallized condition. These differences in behaviour were attributed to the effects of strain-energy stored in the parent-metal grains. The shear and fatigue strengths of joints made in the recrystallized condition were higher than those made in the fine-grained condition. Oxidation tests which were performed at 1000C showed that the oxidation-rate of samples bonded in the fine-grained condition was higher than for those bonded in the recrystallized condition[304]. No localised oxidation of the joint region was observed, indicating that compositional homogeneity existed across the joints. When bonded at 1100 or 1200C, the bonding temperature affected the degree of parent-metal dissolution, the time required to complete isothermal solidification and the extent of microstructural continuity across the joint. Bonding at 1100C did not result in extensive parent-metal dissolution, and shear-testing showed that failure occurred through the center of the joint. Bonding at 1200C increased the degree of parent-metal dissolution and led to appreciable agglomeration of Y_2O_3 particles at the joint interface[305]. Failure then occurred along the joint interface through regions which were depleted in strengthening particles. Bonding at a higher temperature reduced the time required for the completion of isothermal solidification and reduced the amount of strain energy in the alloy, so that grain growth across the joint region was not possible. Bonding was later performed by using 2 to 9μm electrodeposited coatings which were based upon Ni-B or Ni-P interlayers. The coating thickness, and the amount of melting-point depressing boron or phosphorus in the coatings, had an appreciable effect upon microstructural changes in the joint region[306].

MM007

A detailed study was made of the application of transient liquid bonding to superalloys such as MM007, MarM247 and Inconel-713C. They were bonded by using Ni-15Cr-4wt%B amorphous interlayers and IN100 powder sheets with a thickness of about 250μm. In the case of MM007, the times required for the completion of isothermal solidification and the following homogenization process were about 600s and 24h at 1423K, respectively, even for a gap of about 100μm[307]. The rapid homogenization was attributed to the diffusion of elements from the alloying powder to the inter-powder region via three-dimensional diffusion[308]. The tensile strengths of MM007, MarM247 and Inconel-713C joints at 1255K were almost equal to those of the corresponding base metal. It was concluded that it was advantageous to insert alloying powder into wide-gap joints during bonding[309]. The concentrations of aluminium and titanium at the center of bonded interlayers increased and that of chromium decreased with holding time during homogenization; again due to diffusion from the base metal and into the interlayer[310]. Increasing the temperature decreased the holding time required for the completion of homogenization[311]. Dissolution of the base metal into liquid metal during bonding of MM-007 was such that dissolution occurred preferentially at dendrite boundaries and could be described in terms of Nernst-Brunner theory. In the early stages of dissolution, the dendrite boundaries of oxidized base metal were preferentially dissolved and the dissolved zones then linked together. During dissolution, the oxide film separated from the base metal and moved towards the center of the joint interlayer. The dissolution of the oxidized base metal involved three stages, according to the dissolution-rate constant. In stage I, the constant was smaller than that for the base metal without oxidation. Stage II was a transient intermediate stage. In stage III, the constant was the same as that of the base metal without oxidation[312]. The change in the constant was due to the behavior of the oxide film. In stage I, the oxide film was stable and blocked direct contact between the liquid metal and the base metal. In stage II, the oxide film was broken and the liquid metal freely contacted the base metal. The dissolution-rate decreased as the thickness of the oxide film increased. Isothermal solidification of the transient liquid phase during bonding using Ni-B and Ni-P was investigated closely[313]. From experimental data on the isothermal solidification, it was deduced that the process was governed by the diffusion of boron and phosphorus in the base metal[314]. This could be described by:

$$B \text{ in Ni: } D(m^2/s) = 0.14\exp[-226(kJ/mol)/RT]$$

$$P \text{ in Ni: } D(m^2/s) = 0.49\exp[-284(kJ/mol)/RT]$$

The mechanical properties at 1255K of MM007, when bonded using amorphous MBF-30, MBF-80 or MBF-90 interlayers and heating to 1398K, were such that the bonding temperature did not affect the tensile properties of the joint, whereas the bonding and homogenization atmospheres did do so[315]. The liquid-phase content of the alloying powder decreased with bonding time while the boride content at the dendrite boundary in the alloying powder increased with bonding time[316]. It was concluded that boron in the liquid phase was consumed due to boride crystallization in the alloying powder, and that solid-phase growth then occurred in the liquid phase.

Rene N5

The monocrystalline nickel-based superalloy, Rene N5, was bonded by using amorphous nickel-based KNi3A foil and heating at 1513K for 0.25 to 24h under vacuum. The joint region consisted of a bond zone, a diffusion zone and the base metal. The boundaries between them could disappear when the holding time was prolonged, leaving a uniform microstructure[317]. The high-temperature creep rupture properties of the joints could attain 90% of that of the base metal at 1000C under 125MPa. The formation of transjoint grain boundaries and equiaxed grains in the joint resulted in non-monocrystalline joints. During isothermal solidification, the liquid began to solidify via the advance of a flat crystal front growing epitaxially from the unmelted monocrystalline base-metal until the liquid had disappeared[318]. A common single-crystal grain having the same crystal orientation at the interface was then formed. The Rene N5 alloy has also been bonded, by using a Ni-Cr-Co-W-Mo-Ta-Re-B interlayer consisting of plate γ and $M_{23}B_6$ phases, and heating at 1240C for 12h under vacuum followed by post-bonding heat-treatment (1305C, 4h). The molten interlayer alloy exhibited excellent wettability of the nickel-based superalloy. The post-bonding heat-treatment eliminated intermetallic compounds and promoted the formation of γ precipitates in the bond region. It also produced a more uniform microhardness profile across the joint. The shear strength of the joint was also markedly increased, to 533.4MPa, as compared with the value of 437.2MPa without the post-bonding step[319]. When the monocrystalline superalloy was heated at 1473 to 1513K for periods of 0.25 to 12h under vacuum, the temperature/time combination of 1513K and 10h gave the optimum bonding result. The joint, without completion of isothermal solidification, comprised four distinct zones: rapid solidification, isothermal solidification, diffusion and base metal. The isothermal solidification zone across the center-line of the bond had an undifferentiated crystallographic orientation which was the same as that of the base metal. With increasing holding-time at 1513K, the rapid solidification zone and the borides present disappeared; leading to an improvement in the mechanical properties. The high-temperature creep rupture strength (for 100h at 1373K) and the tensile strength (at 1273K) of the joints were equal to 90% of the base-metal values[320]. A nickel-based monocrystalline superalloy has been bonded by using a nickel-based flexible metal cloth as an interlayer and heating at 1473 to 1523K for between 0.5 and 24h under vacuum. The [001] orientation of each specimen was aligned perpendicular to the joint interface. Chemical homogeneity existed across the bond region, and the γ'-phases in the bond interlayer and in the superalloy were almost identical[321]. The bond interlayer almost matched the crystallographic orientation of the bonded substrate. When a nickel-based monocrystalline superalloy was bonded by using a Ni-15Cr-3.5B amorphous ribbon interlayer, the joint consisted of bond, diffusion and base-metal zones, while $M_{23}B_6+\gamma$ and $MB+\gamma$ eutectics formed at the bond center and fine M_3B_2 appeared in the diffusion zone. The γ' phases in the bond interlayer and superalloy substrate were almost identical following homogenization[322]. Due to epitaxial growth of the isothermal solidification fronts from each mating surface, the crystallographic orientation match between the bonded interlayer and the bond substrate was good. When such an alloy was again bonded by using a Ni-Cr-B amorphous foil and heating at 1230C for 8h, and stress rupture tests of the joint and matrix were carried out at 982C and 248MPa or 1010C and 248MPa, the stress-rupture ductility of the joints was appreciably degraded as compared with that of the matrix. This was attributed to solid-solution strengthening by boron, to sub-grain boundaries which formed in the joint zone and to concentrations of creep cavities which appeared in the final stages of stress rupture[323].

Rene 80

This nickel-based superalloy was bonded in the as received cast condition by using a nickel-based interlayer and heating at 1100C under argon or *in vacuo* for various times. The joints were then homogenized (1206C, 1 or 2h). Intermetallics were observed in the joint region, but these were removed by post-bonding heat-treatment (1206C, 2h). The shear strength of a bond which was made *in vacuo* was higher than that of one which was made in argon[324]. The poor shear strength of the latter bonds was attributed to the formation of voids and porosity in the bond region and bond interface during joining.

SXG3

The nickel-based rhenium-containing monocrystalline superalloy was bonded by using a nickel-based hafnium-containing interlayer. Bonding was complete after heating at 1290C in vacuum for 24h. No diffusion-affected zone was observed during bonding. The isothermal solidification stage was accelerated by the precipitation of HfC at the bonding temperature, resulting in a reduced hafnium concentration in the molten zone. It was noted that the interfacial stability of low-angle grain boundaries could be investigated by using this bonding method[325]. The critical misorientation for the discontinuous precipitation of SXG3 along grain boundaries by a hafnium-containing interlayer was between 10 and 17° following heat treatment at 1150C.

A detailed molecular dynamics study has recently been made of the diffusion kinetics of the bonding of nickel-based superalloys using nickel nanoparticles. The calculated self-diffusivity of each element was in a good agreement with experimental data. The slow diffusion led to a more crystalline microstructure, containing nano-sized grains[326]. Non-uniform diffusivity of the surrounding elements affected the diffusion pathways and the joint microstructure.

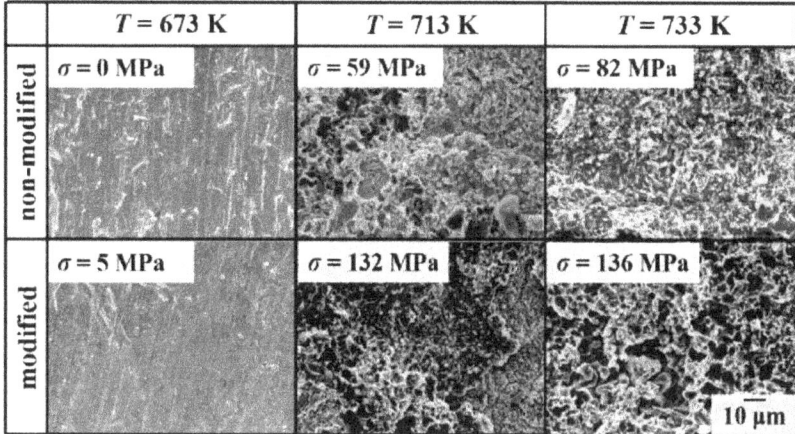

*Figure 20. Scanning electron micrographs of the fracture surfaces of
AC2C|ADC12 aluminium-alloy joints following tensile testing*

Bonding Same-Base Alloys

Copper-Based

AC2C|ADC12

The bonding of AC2C (Al-Si-Mg) to ADC12 (Al-Si-Cu) is complicated by the presence of an oxide film on the zinc interlayer. In order to eliminate this film without recourse to the use of higher temperatures and higher loads, the bonding surfaces were here treated with formic acid and replace the oxide by a metal salt[327]. The tensile strength of the joint increased with increasing bonding temperature, with or without metal salt generation. With metal salt generation processing, high-strength joints were obtained using lower bonding temperatures, as compared with unmodified joints. This was attributed to the generation of metallic zinc via the thermal decomposition of formate within the bond interface at low bonding temperatures. The quality of the joints was assessed largely on the basis of the fracture-surface morphology (figure 20). Using a bonding temperature of 673K, substances did not adhere to either surface. With increasing bonding temperature, substances were observed in pairs on the fracture surfaces. When metal salt generation was used at a bonding temperature of 713K, the fracture surfaces began to exhibit ductile fracture. This was not observed when surface modification was not used.

Cu|Cu-W

Copper has been joined to a Cu-W composite by operating an electron beam welding machine in conduction mode and forming a transient liquid phase at the interface by melting a thin aluminium foil. Under optimum processing conditions, no Cu-Al intermetallics were present in the Cu|Cu-W interface region. Plastic deformation and fracture occurred on the copper side of the joint[328].

Iron-Based

Fe-C|Fe-Cr-Ni

Low-carbon steel has been bonded to stainless steel by using cold-drawing and transient liquid phase diffusion bonding with copper as the interlayer. At the copper|stainless-steel interface, liquid copper diffused along austenite grain boundaries and formed isolated islands of austenite in a copper matrix. A continuous island-like iron-rich phase grew in a pine-tree form at the carbon-steel|copper interface. Within given ranges of temperature and time, the amount of iron-rich phase, the hardness and the shear strength of the bond zone increased with increasing diffusion bonding temperature and time. The compression shear strength of the bond zone was over 300MPa[329].

12Cr1MoV|TP304H

Pipes made from the heat-resistant 12Cr1MoV and TP304H austenitic stainless steels were joined by using an FeNiCrSiB interlayer under an argon atmosphere[330]. The strength of a joint which was made by heating at 1240C for 180s, under a pressure of 4MPa, could attain 590MPa.

45MnMoB|30CrMnSi

These steels were joined via short-term high-temperature heating followed by isothermal solidification at a lower temperature. This two-step heating process could change the interface morphology from planar to non-planar and increase the curvature of the non-planar interface during bonding. The incidence of voids was decreased and the bending strength was increased by increasing the isothermal solidification temperature during the two-step process[331]. Thus, compared with conventional transient liquid phase bonding, the two-step method could reduce the number of voids and improve the bond strength within minutes while using a similar bonding temperature.

SAF2507\AISI304

Transient liquid phase bonding of a super-duplex stainless steel to an austenitic stainless steel has been carried out at 1050C using bonding times of 300s to 0.75h and amorphous Ni-7Cr-4.5Si-3.2B-3Fe alloy (AWSBNi-2) foil as the interlayer. The width of the athermally solidified zone decreased with increasing holding time at a constant temperature. The appearance of hardness peaks in this zone was attributed to the formation of eutectic compounds. The shear strength improved with increasing bonding time. Joints which were produced using a bonding time of 0.75h exhibited the best mechanical behavior, due to the completion of isothermal solidification, and the fracture mode was completely ductile[332]. The stainless steel was bonded to 12Cr-Mo-V-W steel by using iron-based interlayer alloys produced by rapid quenching. Joints made using an Fe-15.7Cr-6Ni-2.8B interlayer possessed microstructures and compositions which were the same as those of the base metal. The mechanical properties were better than those of joints made using the nickel-based interlayer, Ni-15.0Cr-4.0B. Bonds with 12Cr-Mo-V-W steel which were made using an Fe-12.0Cr-3.8B interlayer had almost the same martensite structure as that of the base metal, whereas those made using the nickel-based interlayer remained austenitic after being heated at 1323K for 172.8ks. The hardness and stress-rupture properties were also equal to those of the base metal[333].

T91\Fe-Cr-Mo

This iron-chromium heat-resistant steel has been bonded by using a two-step process which involved short-term heating at a high temperature in order to melt the interlayer, and complete isothermal solidification at a lower temperature[334]. A notable feature was the appearance of a non-planar interface at the join. The solidified bond was deduced to contain a primary solid solution having a similar composition to that of the parent metal, and which was free from precipitates[335]. The tensile strength of the joint was not lower than that of the parent material, and the bend strength was increased due to the non-planar bond. The T91 was bonded to 12Cr2MoWVTiB steel by using FeNiCrSiB as the interlayer. When the bonding pressure was increased from 2 to 6MPa the microstructure and mechanical properties were improved and more similar to those of the base alloys[336]. Theoretical analysis showed that the isothermal solidification completion time could be shortened, because the maximum liquid width decreased under pressure. Pipes made from 12Cr2MoV steel and T91 were joined by using an FeNiCrSiB amorphous interlayer in an argon atmosphere. Pipes made from TP304H and 12Cr2MoV steels were joined by using $Fe_{78}Si_9B_{13}$ and BNi_2 amorphous interlayers. There was a diffusion-asymmetry in the transient liquid phase joining of dissimilar metals, and this was attributed to a difference in the isothermal solidification rate in different materials[337]. This then led to a deviation of the bonding interface, from the original centreline, towards - in the present case, the T91 or TP304H side.

The underlying cause of the asymmetrical diffusion solidification which alters the microstructure during transient liquid phase bonding under a low temperature gradient was studied[338]. A new solute-conserving asymmetrical numerical model, coupled with experiment, showed that a transition from bi-directional to unidirectional solidification, under a constant temperature gradient, is controlled by a competition between liquid and solid-state diffusion at one of the two liquid|solid interfaces.

Nickel-Based

The next level of bonding difficulty might be judged to be the joining of clearly different alloys having the same base. Bonding of a monocrystalline nickel-based superalloy to a polycrystalline nickel-based superalloy for instance has been carried out using diffusion bonding. A 35µm diffusion zone formed via diffusion. In going from the monocrystal to the diffusion zone, the shape of the γ' phase changed from cuboidal to columnar. The hardness of the diffusion zone was slightly higher, and this was attributed to the precipitation of a fine γ'-strengthening phase[339]. The average shear strength was 433MPa and fracture was plastic.

Alloy 247|PWA1483

Columnar-grained Alloy 247 and single-crystal PWA1483, both nickel-based superalloys, have been joined by using transient liquid phase bonding and an amorphous brazing foil which contained boron as a melting-point depressant. At lower brazing temperatures, plate-like and globular boride morphologies developed in both materials. The boride formation occurred in the parent materials at some distance from the solid|liquid interface, and boron-concentration peaks corresponded to their location[340]. It was deduced that the diffusion of another element, which affected the boron solubility was required in order to explain the results[341]. The ratio of the morphologies was temperature-dependent. At very high brazing temperatures, boride formation in the Alloy 247 was avoided; a fact attributed to the three-phase field moving to higher alloying element contents. The formation of borides in PWA1483 could not be entirely avoided by using high brazing temperatures because incipient melting occurred. During subsequent solidification of this area, so-called Chinese-script borides precipitated. The tensile strengths and creep times (to 1% strain) were comparable to, or higher than, those of the weaker joined material for all of the test temperatures and creep conditions which were explored[342].

Materials Research Forum LLC
doi: http://dx.doi.org/10.21741/ 9781644900055

CMSX\Inconel-738

Wide-gap transient liquid phase bonds were made between monocrystalline nickel-based superalloy, CMSX-4, and polycrystalline superalloys, IN738 or IN939, by using the composite interlayers, Niflex-110 or Niflex-115. There was a preferential flow of liquid along the grain boundaries of the Niflex core, rather than a wetting of the facing surfaces, thus leaving non-bonded regions. This could be avoided by increasing the boron content of the Niflex foil, so as to form additional liquid. The suppression of bond-line boride formation was achieved by using wide-gap composite interlayers. The extent of γ' formation at the bond-line, and the presence of second phases in the diffusion zone of the polycrystalline material, were potentially important in controlling the room-temperature mechanical properties of the wide-gap joints. Second-phase formation led to brittle secondary cracks appearing in the polycrystalline substrate[343]. The extent of γ' formation was the predominant factor governing the room-temperature shear strength of the bonds[344]. The bonding of monocrystalline CMSX-4 to polycrystalline IN738 or IN939, using wide-gap composite Niflex-110 and Niflex-115 interlayers as above, was compared with the use of BNi-3 foil. Employment of the composite interlayers suppressed boride formation in the joint[345]. Shear testing revealed that ductile shear failure occurred along the joint and that the extent of γ' formation at the joint was the predominant factor determining the room-temperature shear strength of the joint.

DD98\M963

A study of the bonding of monocrystalline DD98 and polycrystalline M963 nickel-based superalloys showed that, during isothermal solidification at increasing bonding times, the liquid\interlayer thickness decreased. Due to incomplete isothermal solidification, many compounds formed in the centreline. Following bonding at 1190C for 4h, isothermal solidification was complete. An interface formed in the bonding zone due to a crystallographic orientation difference between the parent metals. The stress-rupture life of the joint was greater than 140h at 800C under 350MPa[346]. This was comparable to that of M963. Joint failure occurred in the M963. Monocrystalline DD98 and polycrystalline K465 superalloys were bonded by heating at 1190C for 2h. Numerous phases formed in the center-line of the bond zone due to an incompletely solidified liquid interlayer. Those phases included script-like (chromium-rich CrB boride), tree-like (nickel-rich face-centered cubic $M_{23}B_6$) and blocky compounds (titanium-, tantalum- and niobium-rich MC carbide that resulted from the interdiffusion of carbon between dissimilar base metals) as well as solid-solution γ-phase[347]. Following bonding, many blocky and fine chromium- and tungsten-rich M_6C particles appeared in the diffusion zone, on the K465 side. A number of blocky and platelet-like M_3B_2 borides which were rich in tungsten, chromium and molybdenum precipitated in the diffusion zone on the DD98 side.

Inconel-718|Inconel-625

Inconel-718 and Inconel-625 superalloys were bonded by using the nickel-based alloy, BNi-2, as an interlayer and heating at 1325 to 1394K. The experimentally determined values which were deduced for the activation energy of boron diffusion were very close to those reported for other nickel-based polycrystalline superalloys[348]. On the other hand, the times which were required for complete isothermal solidification were much shorter than those required by other nickel-based superalloys using different nickel-based interlayers. A significant reduction in holding time was needed with increasing bonding temperature and decreasing joint gap.[349] Inconel 718 has been joined to Inconel X-750 by using a Ni-7Cr-3Fe-3.2B-4.5wt%Si foil interlayer and heating at 1373 to 1473K for up to 300s in flowing argon. Good wetting existed between the interlayer and the joined alloys. The microstructures of the joint interfaces exhibited distinct multilayered structures that were due mainly to isothermal solidification and to the following solid-state interdiffusion. Diffusion of boron and silicon from the filler and into the joined metals was the main controlling factor in the microstructural evolution of the joint interface. Silicon and chromium remained in the center of the join and formed brittle eutectic phases[350]. The boron distribution was uniform across the joint area, as it could easily diffuse from the filler and into the base metals.

Inconel-738|Waspaloy

When these alloys were bonded by using Nicrobraz-150, the formation of brittle center-line eutectic micro-constituents was prevented by holding the materials long enough to permit the complete isothermal solidification of molten interlayers in either alloy. The time which was required for that decreased upon increasing the bonding temperature up to 1145C. In joints which were produced at higher temperatures, the formation of non-equilibrium solidification products could not be prevented by holding for the same period which produced isothermal solidification at the lower temperature[351]. In the higher temperature regime, it was also not necessarily possible to prevent the formation of eutectic products in the joint within a reasonable time.

Inconel-738LC|Nimonic 75

The alloys were joined by using a Ni-15Cr-3.5B interlayer and heating at 1080, 1120, 1150C or 1180C for times of 0.5 to 2.5h. At short bonding times, the joint microstructure consisted of continuous eutectic intermetallic phases. Longer times led to an eutectic-free microstructure. For all times and temperatures boride phases were precipitated at the interface between base metal and the interlayer, due to boron diffusion into the base metals[352]. The morphology of the precipitates in the diffusion-affected zone varied from globular to acicular with increasing bonding time. Completion of isothermal solidification prevented the formation of continuous intermetallic phases at the joint center-line[353].

NiAl\MM-247

This intermetallic is important for high-temperature structural applications, but it is essential to be able to bond NiAl to nickel-based alloys. This could be done, as when joining NiAl to itself, by using a BNi-3 melt-spun interlayer. There was precipitation of Ni-Al-based phases and borides within the joint and adjacent regions. There was also martensite formation within the NiAl, and the precipitation of L12-type phases within the joint and at the interface with the NiAl. There was preferential formation of Ni_3B in the nickel, near to the original location of the interface[354]. Polycrystalline and hafnium-doped monocrystalline NiAl alloys could be joined to nickel and nickel-based superalloys by using conventional and wide-gap bonding[355]. Monocrystalline NiAl-Hf was bonded to polycrystalline MM-247 by using a NiAl-Cu composite pre-form interlayer[356]. Monocrystalline NiAl, containing 1.5at% hafnium, has been bonded to polycrystalline nickel-based superalloy, MM247, by using 50μm-thick copper foils and a bonding temperature of 1150C. The hafnium had a marked effect upon the joint microstructure; in particular, the suppression of σ-phase precipitation during bonding[357]. The time required for completion of isothermal solidification was greatly reduced in wide-gap bonds as compared to conventional bonds. Precipitation of the α-phase encountered in polycrystalline-NiAl\MM247 bonds was suppressed during wide-gap bonding of monocrystalline NiAl(Hf) to MM247[358]. The extent of second-phase precipitation in the as-bonded condition was greatly reduced by using the wide-gap technique, but extensive precipitation of HfC and W-rich phases was observed following post-bonding heat-treatment[359]. Single-phase polycrystalline NiAl, multiphase monocrystalline NiAl-Hf and cast polycrystalline MM-247 γ/γ' superalloy were also bonded using conventional 50μm-thick copper interlayers or 175μm composite interlayers involving a NiAl-Cu pre-form. Particular attention was paid to the formation of undesirable second phases such as carbides in the diffusion-affected zone[360,361,362]. The presence of second-phases such as $L1_2$-type γ'-phase – essentially $Ni_3(Al,Ti)$ – as well as MX-type carbides, σ-phase intermetallics and elemental chromium and tungsten were correlated with mechanical properties such as the room-temperature shear strength. The formation of such undesirable second phases could be a major difficulty for the transient liquid phase bonding of NiAl to nickel-based superalloys. The formation of those second phases does not appear to be associated directly with the interlayer used. The formation of second phases in the bonds instead seems to result from extensive interdiffusion between the two materials, combined with a low solubility - in NiAl - of common alloying additions such as chromium, molybdenum and tungsten. Use of a composite interlayer which comprised non-melted NiAl and molten copper permitted the choice of very short bonding times; a factor which is known to suppress the formation of σ-phases and carbides. On the other hand, the use of NiAl-Cu composite interlayers had little effect upon the formation of second phases during post-bonding heat treatments. Investigations have thus been made of the use of Ni_3Al-Cu interlayers, as the latter can prevent the entry of chromium, molybdenum and tungsten into the NiAl; thus suppressing the formation of σ-phase, for example, during post-bonding heat exposure. Bonding of Ni-24Al-16atCr and Ni-30at%Al having a tetragonal L10-type martensitic microstructure, to stoichiometric NiTi having a monoclinic distorted B19-type martensitic matrix, was carried out by using a 50μm-thick copper interlayer and heating at 1150C for between 1200s and 2h. A holding time of 1h at 1150C was sufficient to remove liquid phase from the bond-line, although a layer of Ni_2AlTi L2- type Heusler phase was left at the bond line. The latter remained after further holding of up to 2h at 1150C. When using a 50μm-thick copper interlayer, bonding for 2h at 1150C led to appreciable titanium diffusion from the NiTi substrate and into the joint[363]. Depletion of titanium from the NiTi then reduced the solidus temperature of the NiTi to a point where localized melting occurred on the surface of the NiTi. The NiAl was largely unchanged during bonding. The resultant bonds were post-bonding heat-treated at 1000C for between 4 and 12h. There was formation of Ni_2AlTi (L21 structure) at the bond-line and differences in the mode of Ni_2AlTi precipitation between the bonding and post-bonding heat treatment[364]. There were changes in the nature of $Ni_3(Al,Ti)$ precipitation, in the NiTi substrate, from DO24 to L12 type.

Titanium-Based

Ti|Ti-6Al-4V

Commercially pure titanium and Ti-6Al-4V have been bonded by using silver or copper interlayers. Intermetallics always formed in the joint area when a silver interlayer was used, and copper was thus a more viable candidate for joining titanium. At 1000C or above, no intermetallics were found in the bond region of the purer material. The latter region consisted of eutectic mixtures and Ti-Cu solid solutions. The microstructure of the Ti-6Al-4V joint center-line consisted of Cu-Ti eutectic and Ti-Cu solid solution plus other alloying elements[365]. The maximum strength obtained for the purer material, using a copper interlayer, was 470MPa.

$TiAl|Ti_3Al$

Transient liquid-phase diffusion bonding of a Ti_3Al-based alloy to TiAl has been carried out by using a Ti-13Zr-21Cu-9wt%Ni interlayer foil. The $Ti_3Al|TiAl$ joint consisted mainly of a titanium-rich phase, a Ti_2Al layer, an α_2-Ti_3Al band and residual interlayer alloy. The amount of residual interlayer in the central part of the joint decreased with increasing bonding temperature, while the Ti_2Al and α_2-Ti_3Al reaction bands close to the Ti_3Al-based alloy gradually thickened. The central part of the joint exhibited the greatest microhardness across the entire joint. Joints which were bonded at 1193K for 600s using a pressure of 2MPa (figure 21) exhibited the highest shear strengths: 417MPa at room temperature and 234MPa at 773K[366]. Similar bonding of Ti_3Al-based alloy to TiAl, but using a Ti-Zr-Cu-Ni-Fe alloy interlayer, produced joints which consisted mainly of titanium-rich phase, a Ti_2Al layer, an α_2-Ti_3Al phase plus residual aluminium-dissolved interlayer alloy. Shear test results here indicated that, between 880 and 1010C, an increase in bonding temperature improved the joint strength.

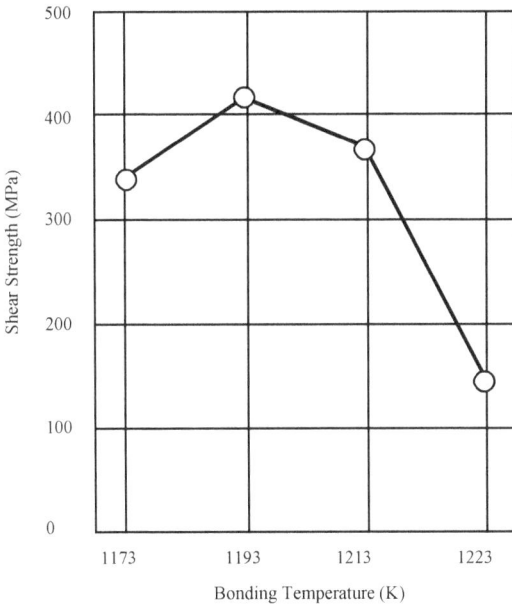

Figure 21. Effect of bonding temperature on room-temperature shear strength of $Ti_3Al/TiAl$ joints (bonding pressure: 2MPa; dwell time: 600s).

The maximum shear strength of the $Ti_3Al|TiAl$ joint was 502MPa at room temperature and 196MPa at 500C[367]. Finally, the use of a Ti-15Cu-15wt%Ni interlayer led to the appearance of a Ti_2Al layer, an α_2-Ti_3Al band, a titanium-rich phase, Ti_2Ni and Ti_2Cu in the joint. The latter had a shear strength of 74 to 79MPa. The use of a new Ti-Zr-Cu-Ni-Co filler alloy led to joints which consisted mainly of titanium-rich phase, a Ti_2Al layer, a $(Ti,Zr)_2Al$ phase plus $(Ti,Zr)_3Al$ with copper, nickel and cobalt. Joints which were bonded at 1253K for 600s using a pressure of 2MPa had a shear strength of 435MPa at room temperature and 234MPa at 873K. Here, CuZr and Zr_2Ni rather than Ti_2Ni and Ti_2Cu were found in the joint. The formation of the CuZr and Zr_2Ni, as well as an α-titanium phase greatly increased the joint ductility[368].

TiAl|Ti6242

The TiAl and Ti6242 were bonded by using Ti-Cu foils as interlayers. The bonding pressure played an important role even though a liquid phase existed during bonding. The stacking sequence of the parent materials also had a large effect upon joint formation. Elimination of defects from the TiAl/interlayer interface was a major problem[369]. Similar interface structures were observed in all cases, but the thickness of the joint zone varied. Surface oxide layers on the TiAl also had a large influence on interface diffusion and a so-called bridge effect was observed. The diffusion copper atoms in TiAl was controlled by vacancies beneath the surface[370].

Ti-6Al-4V|TiNi

The bonding of nickel and titanium, described elsewhere, was potentially relevant to the reaction-assisted diffusion bonding of TiNi to Ti-6Al-4V. This was carried out using Ni/Ti multilayer foils produced by roll-bonding or Ni/Ti multilayer thin films prepared by magnetron sputtering. In the former case, nanostructures were produced by severe plastic deformation (strain-rate of 20/s for up to 16 cycles) of stacked alternating nickel and titanium foils, giving a total thickness close to 300µm. Using these multilayers, it was possible to produce joints by solid state joining (800C, 1h, 10MPa) having a shear strength of 35MPa. The layered structure of the foil transforms into TiNi, with small regions of Ti_2Ni and $TiNi_3$, while a continuous layer of Ti_2Ni was present close to the Ti_6Al_4V. Nickel/titanium multilayer thin films with nanometre modulation were also used to produce (750 or 800C, 1h, 10 or 50MPa) joints having a shear strength of 88MPa[371]. The interface was less than 10µm and comprised several zones, including a Ti_2Ni layer close to the Ti_6Al_4V, followed by TiNi and Ti_2Ni. Solid-state diffusion bonding (750, 800 or 900C, 1h, 10MPa) of TiNi to Ti-6Al-4V had previously been performed using reactive nickel/titanium multilayer thin films. The surfaces were modified by sputtering alternating nickel and titanium nano-layers in order to increase the diffusivity at the interface. Joints which were free of porosity and cracks were produced using nickel/titanium reactive multilayer thin films. The reaction zone constituted of columnar grains of Ti_2Ni and $AlNi_2Ti$ close to the Ti_6Al_4V, and of alternating layers of Ti_2Ni and TiNi equiaxed grains. The grain size decreased in going from the Ti_6Al_4V to the TiNi, and nanometre grains were observed in layers closest to the TiNi base material[372].

Bonding Different-Base Alloys

The next logical 'rung on the ladder' is the joining of alloys having completely different elements as their base. This requirement is usually imposed by the frequent common use of key alloys in the same type of application: a typical and important case being the aerospace industry.

Aluminium|Iron

AA6005|S355

The aluminium alloy was bonded to the steel by using a pure copper foil interlayer. At the beginning of the process, the foil produced an eutectic liquid which wetted the surfaces of the steel and aluminium alloy and blocked interdiffusion between iron and aluminium in the later stages of the process. After heating at 580C for 0.25h, a layered structure appeared in the joint. With increasing holding time and temperature, the width of the diffusion zone increased and the layered structure was no longer obvious[373]. Cracks appeared in the Fe-Al intermetallic compound layer near to the steel side, and the microhardness attained 439HV. At a time and temperature of 0.5h and 580C, the shear strength of the joint attained its highest value of 77MPa. Aluminium-containing magnesium alloys and steels have been joined by using a silver interlayer. A magnesium-silver eutectic melt was produced at 773K, and a uniform nanoscale Fe-Al reaction layer formed at the melt|steel interface during isothermal solidification of the melt, driven by diffusion of silver into the magnesium alloy[374]. Following solidification, the bond yield strength exceeded that of the original magnesium alloy.

An aluminium metal-matrix composite was joined to a low-carbon steel by using a 50μm copper foil and heating at 590C for 0.5 or 2h. The residual interlayer|composite interface was weaker than the interlayer|steel interface. The edge of the composite was preferentially dissolved by extruded liquid and this led to the appearance of a circular notch around the contacting surface. The extruded liquid phase spread towards the composite side rather than towards the steel side[375]. Under low pressures, joint fracture occurred within the residual interlayer. Under higher pressures, it occurred at the surface of the composite. Iron could be copiously dissolved in the interlayer and form intermetallic compounds in the metal matrix, thus leading to an impaired joint strength.

Aluminium|Magnesium

LY12|AZ31B

The alloys were bonded by using a copper foil interlayer and heating under vacuum at 450C for 1h under a pressure of 2.5MPa. The microstructure in the interface zone of the diffusion joint included an α-magnesium solid solution, Mg_2Cu, Cu_2Mg, γ-copper solid solution, CuAl, $CuAl_2$ and β-aluminium solid solution. The width of the interface increased with increasing bonding temperature, and the microhardness gradient in the interface zone on the aluminium alloy side was much steeper than that on the magnesium alloy side. As the joining temperature was increased, the shear strength first increased and then decreased[376]. Joining conditions of 460C and 2.5MPa led to the highest joint shear strength of 35MPa. When the bonding of AZ31 and AA5083 was performed using impact pressure under vacuum, the joints comprised four layers: magnesium-alloy matrix, reaction layer, diffusion layer and aluminium-alloy matrix. There were also the intermetallics, Mg_2Al_3, MgAl and $Al_{56}Mg_{44}$. The greatest hardness, of 3300MPa, was found in the joint zone[377]. The tensile strength of the joint first increased and then decreased, with a maximum of up to 46MPa. The fracture surface was a mixture of quasi-cleavage and dimple. These alloys were also bonded by using zinc as an interlayer. As the holding time at low temperature was increased, the diffusion layer thickness and the tensile strength increased, attaining 38.5MPa. The failure mode was brittle[378]. Here, $MgZn_2$ plus a little $Mg_{17}Al_{12}$ were found and the maximum hardness was 170VHN. They were additionally bonded by using a so-called supercooled process with no middle layer. The tensile strength of the joint increased with increasing treatment time and could attain 20.5MPa. Cleavage planes and tearing ridges, usually taken to be indicators of cleavage fracture and intergranular fracture, could be seen on the tensile fracture surfaces on the aluminium alloy side. The failure mode on the magnesium alloy side was typical intergranular fracture[379]. A layer of intermetallic compound formed between the magnesium and aluminium alloys, leading to a low strength in the joint region with MgAl, Mg_2Al_3, $Mg_{44}Al_{56}$ and $Mg_{17}Al_{12}$ intermetallic compounds. The greatest hardness in the joint zone was 320VHN. The AZ31B alloy was joined to copper by heating at 500C for 0.67h under a pressure of 2.5MPa. A diffusion interface zone with a width of some 450μm was formed. The microstructure of the joint included a grain-boundary penetration layer which comprised α-Mg and $Mg_{17}(Cu,Al)_{12}$ which precipitated along the grain boundaries of α-Mg solid solution, plus an eutectic of α-Mg with Mg_2Cu, the intermetallic compound, Cu_2Mg, an eutectic of α-Mg with Mg_2Cu, and Cu(Mg) solid solution. The interface width increased with increasing holding time, and the width of the eutectic beside the Cu_2Mg layer in the interface zone clearly increased[380]. The microhardness of the interface zone was much higher than that of the magnesium alloy or the copper, and there were 4 distinct regions of hardness. The microhardness of the interface increased with increasing holding time. Diffusion bonding was used to join AZ31B magnesium alloy to copper, with a nickel foil interlayer. Good joints, with magnesium grain-boundary penetration, were obtained at 500C after 0.33h. When using an intermittent gradient pressure, the join zone reached its greatest width (0.18mm) and the microstructure of the join zone comprised $Cu_{11}Mg_{10}Ni_9$, an α-Mg-Mg_2Cu-Mg_2Ni) eutectic structure and α-Mg. Under gradient pressure conditions, the width of the join was reduced and its microstructure comprised mainly α-Mg-Mg_2Cu-Mg_2Ni eutectic structure and $Cu_{11}Mg_{10}Ni_9$ compounds. When using a constant pressure, the join microstructure was similar to that under intermittent gradient pressure conditions. The join here took its smallest width (0.11mm), and the size of the α-Mg grains was reduced[381].

The kinetics of isothermal solidification during the partial transient liquid phase bonding of aluminium-Mg_2Si composite to the magnesium alloy, AZ91D, were such that, upon increasing the temperature, the magnesium content in the activated bond surface decreased and isothermal solidification occurred at the mating surface of the base metals[382]. The amount of magnesium increased when the activity of the interface was increased by imposing a higher temperature gradient. These materials have also been bonded by using two different heating-rates. Upon decreasing the heating rate from 20 to 2C/min, the magnesium content in the joint line decreased and the microstructure changed. A heating rate of 2C/min resulted in an increased shear strength of the joint. The kinetics of bonding process were accelerated due to an increase in solute diffusivity through the grain boundaries of the metal-matrix composite[383]. It was suggested that magnesium and silicon in the interlayer promoted the partial disruption of oxide films and thus facilitated bonding.

Aluminium|Titanium

AA2024|Ti-6Al-4V

It has been found to be of particular interest to join Ti-6Al-4V to a range of other alloys. Transient liquid phase bonding of Al2024 and Ti-6Al-4V was achieved by using a Cu-22%Zn interlayer at 510C under vacuum (0.01Pa). Joint formation was attributed to the solid-state diffusion of copper and zinc into the Ti-6Al-4V and Al2024, followed by eutectic formation and isothermal solidification along the Cu-Zn|Al2024 interface. The joint hardness at the interface increased with increasing bonding time and was attributed to the formation of intermetallics such as Al_2Cu, $TiCu_3$, $Al_{4.2}Cu_{3.2}Zn_{0.7}$, $Al_{0.71}Zn_{0.29}$,

Ti_2Cu, $TiAl_3$ and $TiZn_{16}$ in the joint zone[384]. The shear strength of the joint was a maximum of 37MPa for a bonding time of 1h. Those two alloys have also been joined at 453C using a 50μm-thick Sn-5.3Ag-4.2Bi interlayer. Examination revealed the presence of Ag_3Al, Ag_2Al and Al_3Ti, plus tin as a solid-solution bond. The bismuth was homogeneously distributed in the two base metals. The tensile strength of the joint was 62MPa.[385] Joining using a Sn-Ag-Cu-Ni interlayer, with a thicknesses of 40, 80 or 120μm, was performed at 510C under a pressure of 10^{-4}mBar. With increasing bonding time, various intermetallic compounds formed in the joint zone. Diffusion generally led to isothermal solidification and process completion within the 1h bonding time, but remains of the 120μm-thick interlayer were found. With increasing bonding time, the hardness of the joint increased to 139VHN[386]. The shear strength was proportional to the bonding time, but there was an optimum interlayer thickness for achieving maximum shear strength.

AA7075|Ti-6Al-4V

These aerospace alloys have been bonded at 500C, using 22μm-thick copper interlayers and various bonding times. Joint formation was attributed to the solid-state diffusion of copper into the titanium alloy and into Al7075, followed by eutectic formation and isothermal solidification along the Cu|Al7075 interface. The joint region revealed the formation of eutectic phases such as Al_2Cu, $Al_2Mg_3Zn_3$ and $Al_{13}Fe$ along the grain boundaries within the Al7075 matrix. At the Cu|Ti alloy bond interface, a solid-state bond formed; leading to Cu_3Ti_2 phase formation along that interface. The joint region homogenized with increasing bonding time and exhibited a maximum bond strength of 19.5MPa for a bonding time of 0.5h[387]. These same alloys were bonded at 500C, under a pressure of 5 x 10^{-4}torr, after electrodepositing copper onto the contacting surfaces. A film of 50μm-thick Sn-4Ag-3.5Bi was used as an interlayer. Eutectic and intermetallic (θ-Al_2Cu, TiAl, Ti_3Al) formation along the Al7075 grain boundaries and Ti|Al interface led to joint formation at the aluminium and titanium interfaces. The hardness of the joints increased with increasing bonding time, and this was attributed to intermetallic formation at the interface[388]. The highest bond strength of 36MPa was found the samples which had been joined for 1h.

Cobalt|Nickel

FSX-414|GTD-111

The nickel-based GTD-111 was joined to cobalt-based FSX-414 by using a 50μm-thick amorphous Ni–Si–B interlayer. The athermally solidified zone in the middle of the joint contained a ternary γ|Ni_3B|Ni_6Si_2B eutectic plus a little Ni_3Si. The amount of eutectic

Materials Research Forum LLC
doi: http://dx.doi.org/10.21741/ 9781644900055

phase present decreased with increasing bonding time. Complete isothermal solidification had occurred after 100min, and this process was governed by the diffusion of silicon and boron from the bonding zone into the base metals, by the diffusion of titanium, aluminium, cobalt and chromium from the base metals into the bonding zone and by their proportional re-distribution within the joint. The width, of 80μm, of the joint zone was greater than the thickness of the interlayer, due to coupled diffusion and dissolution. Following homogenization heat-treatment, there was a 34% increase in the shear strength of the joint, due to a uniform distribution of strengthening precipitates within the joint[389]. While bonding GTD-111 to FSX-414, it has been noticed that the amount of γ/nickel boride/Ni-B-Si ternary eutectic in the athermally solidified zone of joints increases with increasing thickness of the interlayer. Upon increasing the temperature from 1130 to 1160C, complete isothermal solidification was achieved and no harmful phases were found at the joint centreline. At higher temperatures, dissolution of the substrate elements blocked the diffusion of melting-point depressant elements, and the eutectic phase reappeared[390]. Solidification cracking occurred when the bonding was not performed *in vacuo*. A hardness peak persisted in the diffusion-affected zone even when the isothermal solidification was complete. A maximum shear strength of 412MPa was found following complete isothermal solidification of a bond made using a 50μm interlayer and heating at 1160C for 0.75h. A study of the transient liquid phase bonding of nickel-based superalloy and cobalt alloy suggested that, contrary to the assumption that rapid atomic diffusion in the liquid would reduce the processing time required to eliminate completely any liquid phase from the joint region by diffusional solidification, the occurrence of liquid-state diffusion would markedly prolong the processing time (figure 22)[391]. A theoretical prediction of the effect of liquid-state diffusion on the joint microstructure which is developed during the transient liquid phase bonding of dissimilar superalloys, and which was based upon an asymmetrical numerical simulation model, has been experimentally verified. This led to an improved understanding of the diffusional solidification mechanism during transient liquid phase bonding, and permitted the creation of a monocrystalline joint microstructure between dissimilar polycrystalline and monocrystalline substrates[392]. This had been deemed impossible by conventional theoretical models for transient liquid phase bonding.

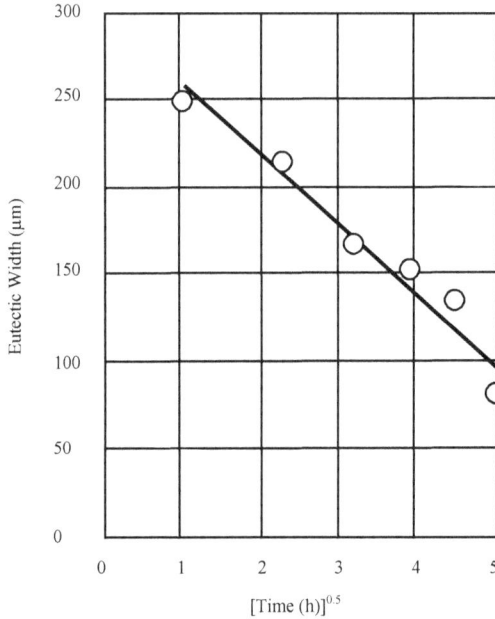

*Figure 22. Average eutectic width in an IN738/cobalt joint as a
function of the square root of the holding time for an initial gap of 200μm*

Copper|Iron

Cu|Fe-Cr-Ni

Vacuum diffusion bonding of stainless steel to copper has been performed (830 to 950C, 1h, axial pressure: 3MPa) using tin-bronze, gold or tin-bronze/gold composite foil interlayers. Grain-boundary wetting occurred within the steel adjacent to the interface, due to contact melting between the tin-bronze and gold when the composite interlayer was used. This wetting could already occur at 830C, and became significant with increasing temperature[393]. The tensile strength of joints made using the composite was greater than that for joints made using the separate interlayers, and reached 228MPa for a joining temperature of 850C. Austenitic stainless steel and copper were diffusion-bonded

(875C, 0.5h, pressure: 3MPa) either conventionally or with an added external electrical current. The greatest hardness occurred in regions next to the interface, and gradually decreased away from the interface. The highest interface strength was obtained when an external current was applied; those specimens exhibited a higher degree of diffusion[394].

B10|Fe-C

The Cu-Ni alloy, B10, was joined to low-carbon steel by using an H62 brass interlayer, and a good bond was obtained by heating at 950 to 1000C for 0.5 to 2h. The composition changed gradually at the brass|Cu-Ni interface, and the grains grew continuously. There was no marked change in hardness. Large variations in composition were found at the steel|brass interface. A pearlite-rich band lay, parallel to the bonding line, in the region of steel next to the interface[395]. When the time and/or temperature was increased, formation of an iron-rich microstructure having a greater hardness led to a decrease in the plasticity of the interface.

Cu-Al-Be| 1Cr18Ni9Ti

When the copper alloy was bonded to the stainless steel by using Cu-30wt%Mn alloy as an interlayer and heating at 1223K for 40min under 1MPa, the joint strength could attain 487MPa and the fracture was plastic[396]. The structure of the interface and the strength of the joint were greatly improved by an increased manganese content[397].

Copper|Titanium

Cu|Ti-6Al-4V

Diffusion bonding of Ti-6Al-4V and oxygen-free copper has been carried out at low temperatures by using a 30μm silver foil interlayer. The latter prevented the formation of titanium-copper compounds, while AgTi and $AgTi_2$ appeared at the Ti-6Al-4V|silver interface and a solid solution formed at the silver|copper interface. The tensile strength of the joint first increased and then decreased with increasing bonding temperature or time. A joint which had been prepared at 700C, using a time of 2h and a pressure of 10MPa, had a maximum tensile strength of 161.9MPa and ductile fracture nucleation and growth occurred via the silver|copper interface[398].

Copper|Tungsten

Cu|W

Tungsten was joined to a copper alloy by using Cu-33Mn or Cu-33Mn-5wt%Ni as the interlayer. The microstructure of joints made using the Cu-33Mn-5Ni interlayer was denser than that for joints made using the Cu-33Mn interlayer[399]. The addition of nickel increased the diffusion rate, the isothermal solidification rate and the degree of Ni(W) solid-solution formation.

Gold|Nickel

Au-Sn|Ni-Cu

Transient liquid phase bonded Au-Sn|Ni(Cu) joints have been found to possess excellent mechanical properties even after exposure to 400 or 450C, especially the Au-Sn|Ni-30Cu and Au-Sn|Ni-40Cu systems. The shear strength could attain 75MPa in the case of Au-Sn|Ni-40Cu joints, and this increased to 87 and 97MPa after exposure to 400 or 450C for 24h, respectively. During bonding at 350C, ordered AuCu, Ni_3Sn_2 and $\alpha(Au)$ formed in the Au-Sn|Ni-Cu joints. The ordered AuCu phase then transformed into disordered (Au,Cu), and a new Ni_3Sn phase appeared between the substrate (Au,Cu) and the Ni_3Sn_2 layers during exposure at 400 or 450C. In the Au–Sn|Ni system, only a Ni_3Sn_2 layer formed during bonding and a new Ni_3Sn layer appeared near to the substrate during exposure at 400 or 450C. The ductile ordered AuCu or disordered (Au,Cu) layers relieved local stress concentrations in the joint under shear stress and thus improved the mechanical reliability. The maximum stress in Au-Sn|Ni(Cu) joints decreased with increasing thickness of the Au-Cu layer. The Au–Sn|Ni joint consequently exhibited a lower shear strength as brittle Ni_3Sn or Ni_3Sn_2 layers formed near to the substrate[400]. The shear strength of Au–Sn|Ni–Cu joints increased for thicker Au–Cu layers.

Iron|Magnesium

AISI304|AZ31B

Transient liquid phase bonding of the magnesium alloy, AZ31B, and AISI304 stainless steel was carried out by using a copper interlayer. When the joining conditions were 510C for 0.5h or 530C for 600s, no eutectic phase was found in the joint and the metallurgical bonding was weak. When the joining conditions were 520C for 0.5h or 530C for 0.33h, the joint interface contained Mg-Cu eutectic. The joint width significantly increased and the interfacial bonding strength improved. The choice of 530C and 0.5h produced a laminar diffusion region with a width of about 350µm on the magnesium side of the joint. The microstructure consisted of a Mg-Cu eutectic layer followed by a magnesium-rich solid solution layer, with $Mg_{17}(Cu,Al)_{12}$ dispersed in a magnesium alloy matrix and zones with Mg-Cu-Al ternary compounds in the grain boundaries of magnesium alloy. The maximum shear strength of the joint was 52MPa. When using bonding conditions of 540C for 0.5h or 530C for 0.67h, isothermal solidification occurred in eutectic liquid of

Materials Research Forum LLC
doi: http://dx.doi.org/10.21741/ 9781644900055

the interfacial diffusion zone and the joint shear strength decreased[401]. This was attributed to the presence of a continuous network of Mg-Cu-Al compounds along the magnesium grain boundary. Titanium and AISI304 stainless steel have been joined by using a 0.1mm-thick aluminium foil interlayer at 550, 650 or 700C for 1h under a 2MPa pressure in vacuum. Intermetallic layers of $FeA1_3$ and Fe_2Al_5 were detected at the stainless-steel|aluminium interfaces. Intermetallic layers of TiAl, $TiAl_2$ and $TiAl_3$ were found at the aluminium|titanium interfaces. The width of the intermetallic layers of both interfaces gradually increased with increasing bonding temperature. Irregular Al_7Cr particles were also observed in the aluminium matrix for joints which had been transient liquid phase bonded at 700C[402]. The microhardness of the joints attained values of 220 to 870HV; higher than that of the base metals. Transient liquid phase diffusion bonding of TiNi shape memory alloy to SUS304 stainless steel was performed using AgCu foil as an interlayer[403]. The maximum shear strength of the joint occurred for specimens which had been joined at 860C during 1h under a pressure of 0.05MPa[404]. The joint's properties were appreciably affected by the formation of $TiNi_2$, TiFe and Ti_3Ni_4 etc. TC4 titanium alloy and AISI304 were bonded by using a Cu-Ti-Ni-Zr-V amorphous interlayer. Good joints could be obtained due to an excellent compatibility between the interlayer metal and the stainless steel substrate. Partial dissolution of the latter occurred during bonding. The shear strength of a joint made using $Cu_{43.75}Ti_{37.5}Ni_{6.25}Zr_{6.25}V_{6.25}$ foil was 105MPa and that of one made using $Cu_{37.5}Ti_{25}Ni_{12.5}Zr_{12.5}V_{12.5}$ was 116MPa. The joints fractured through the joint and exhibited brittle fracture features. The intermetallics, Cr_4Ti, $Cu_{0.8}FeTi$, Fe_8TiZr_3 and Al_2NiTi_3C were found in fractured joints made using $Cu_{43.75}Ti_{37.5}Ni_{6.25}Zr_{6.25}V_{6.25}$ foil, while Fe_2Ti, TiCu, Fe_8TiZr_3 and $NiTi_{0.8}Zr_{0.3}$ were detected in a joint made using $Cu_{37.5}Ti_{25}Ni_{12.5}Zr_{12.5}V_{12.5}$ foil. The presence of Cr-Ti, Fe-Ti, Cu-Fe-Ti and Fe-Ti-V intermetallics in the joint seam provoked their fracture[405]. The stainless steel and Zr-Sn-Nb alloy were joined by means of partial transient liquid phase bonding using a nickel interlayer. Reaction layers were formed in both joints and comprised an σ-FeCr layer, a $Zr(Cr,Fe)_2+α$-Zr layer and an α-Zr+Zr_2(Ni,Fe) layer. The intermetallics were relatively compact, and cracks formed in the reaction layer of a direct-bonded joint. In joints made using a nickel interlayer, many dispersed α-Zr phases existed in the reaction layer, and the thickness of the latter layer was clearly greater than that without a nickel interlayer[406]. Due to the lower residual stresses and wider crack-free reaction layer, the strength of the joint was increased by using a nickel interlayer.

Iron|Nickel

GH3039|IC10

Monocrystalline Ni$_3$Al-based IC10 was bonded to the stainless steel by using a BNi-2 interlayer and heating at 1050 to 1200C for 2h. Precipitates in the diffusion zone, on the GH3039 side, gradually decreased in number and disappeared with increasing bonding temperature. A layer of borides of great hardness, in joints made at 1150C markedly reduced the room-temperature tensile strength but had little effect upon the high-temperature tensile strength of the joint[407]. The boride layer was the preferred crack path during both room-temperature and high-temperature tensile testing. Joints with dispersed precipitates had the highest room-temperature tensile strength and the homogenized joints exhibited the highest ductility.

Iron|Titanium

Fe-C|Ti

Commercial-purity titanium and low-carbon steel have been joined at 870 or 910C by using a copper interlayer. Zones having various chemical compositions formed in the joint region, and points of steel|copper contact exhibited the greatest hardness[408]. This was due to the formation of FeTi and Ti$_3$Cu$_4$ at the interface. When bonding titanium to some other steels using a copper-based interlayer, no joining could be achieved below 800C even after holding for 3h. Bonding was possible at 850C for all holding times. Atom diffusion and migration between titanium and iron or carbon were effectively prevented by the copper-based interlayer and neither Fe-Ti nor Ti-C intermetallics formed in the joint[409]. In general, the direct joining of titanium alloy to stainless steel leads to the formation of weakening Fe-Ti and Fe-Cr-Ti intermetallics in the diffusion zone due to the limited mutual solid solubilities of iron, chromium, nickel and titanium.

Fe-Cr-Ni|TiNi

The shape memory alloy was bonded to stainless steel by using a Ag-Cu foil interlayer. The maximum joint shear strength was found after heating at 860C for 1h under a pressure of 0.05MPa[410]. These properties were markedly affected by the formation of TiNi$_2$, TiFe, Ti$_3$Ni$_4$, etc. The microhardness of the diffusion zone on both the TiNi and stainless-steel sides ranged from 500 to 650VHN. The microhardness of the middle zone was only about 120VHN[411]. The interface of the joint consisted of a TiNi transition zone, a middle zone and a stainless steel transition zone. The main phases present therein were Ti(Cu,Ni,Fe), AgCu and TiFe, respectively. With increasing temperature or holding time,

Materials Research Forum LLC
doi: http://dx.doi.org/10.21741/ 9781644900055

the shear strength of the joint first increased and then decreased[412,413]. The highest shear strength was about 239.4MPa, and fracture occurred in the diffusion interface between the TiNi and the AgCu interlayer via mixed ductile and brittle failure. The corrosion behavior of the joint in Hanks solution at 37C showed that its corrosion resistance was comparable to, but lower than, that of the base metals during early anodic polarization. The corrosion rate of the joint was between that of the TiNi and stainless-steel in the transpassive region at high potentials. Both materials exhibited localized corrosion, plus a little pitting corrosion[414]. The joint exhibited mainly pitting corrosion which was concentrated at sites of higher copper concentration.

AISI304|Ti-6Al-4V

During the diffusion-bonding of Ti-6Al-4V and the stainless steel, atomic diffusion between titanium and iron or carbon is prevented by using a pure silver interlayer. The shear strengths of joints are directly related to the bonding time, and a maximum lap shear strength of 149MPa and 18% failure strain have been obtained for a bonding time of 1h[415]. Samples of Ti–6Al–4V and AISI304 have been joined by using a copper interlayer at 870 to 960C. The width of the joint first increased and then decreased with increasing bonding time. The occurrence of intermetallics such as Ti_2Cu, Ti_2Cu_3, $TiCu_2$, $TiCu_4$, TiCu and FeTi was expected for various bonding times, except 1h. Lengthening the holding time from 0.25 to 1h resulted in an isothermally solidified joint, with single-phase solid-solution β-Ti at the interface. The rate of isothermal solidification increased with increasing bonding temperature and, when the bonding temperature was increased to 960C, the width of the eutectic and intermetallic zones was eliminated, leading to complete isothermal solidification. The maximum shear strength of 374MPa was also found for a bonding time of 1h, and this amounted to 70% of the strength of Ti–6Al–4V. A minimum hardness of 216VHN was found for the joint zones. Fracture involved a mixture of failure modes: brittle and mixed brittle-ductile.[416,417].

AISI310|TiAl

Joining of TiAl to stainless steel has been achieved by using reaction-assisted diffusion bonding (700 or 800C, 1h, compressive stress: 10, 25 or 50MPa), with 14nm alternating direct-current magnetron-sputtered nickel and aluminium nano-layers having a bilayer thickness of 30 or 60nm. The bonding interfaces were some 5μm thick, with a layered microstructure. The interface was composed mainly of equiaxed grains of NiTi and $NiTi_2$. The use of the multilayers led to joints having lower hardness values, closer to those of the joined materials, together with higher shear strengths[418]. The highest shear

Materials Research Forum LLC

doi: http://dx.doi.org/10.21741/ 9781644900055

strength was obtained for the joint produced (800C, 1h, 10MPa) using nickel/titanium multilayers with a 30nm bilayer thickness[419].

AISI316L|Ti-6Al-4V

The stainless steel and Ti-6Al-4V were here joined by using pure copper interlayers of various thicknesses. Increasing the interlayer thickness led to incomplete bonding within 1h. A maximum shear strength of 220MPa was found for a bonding temperature of 900C. Upon increasing the bonding temperature to 960C, there was a decrease in the bond strength[420]. This was attributed to an increase in the width of the joint zone and to the formation of more brittle intermetallics at the interface.

UNS32750|Ti-6Al-4V

Joining of Ti-6Al-4V and the super duplex stainless steel was carried out by using a copper interlayer, at 890, 920, 950 or 980C, and a bonding time of 1h. The isothermal solidification zone increased, with increasing temperature, up to 980C with the formation of TiCu, Ti_2Cu, Cr_2Ti, Fe_2Ti and FeTi. Increasing the bonding temperature promoted their formation by increasing the diffusion of titanium, copper and iron, and thus decreased the shear strength of the joint. The joint made at 890C had the highest shear strength of 271MPa; amounting to 49% of the strength of Ti-6Al-4 V and 47% of that of UNS32750. The average hardness of the bond was also better than that of either base metal[421]. The stainless-steel side had hardness values ranging from 420 to 460HV, while the other side had hardness values ranging from 330 to 370HV. Diffusion-bonding (875 to 950C, 1h, vacuum, 4MPa uniaxial pressure) was used to join duplex stainless steel to Ti-6Al-4V, with both nickel and copper interlayers. After 875C joining, layered Cu_4Ti, Cu_2Ti, Cu_4Ti_3, CuTi and $CuTi_2$ were present at the copper|Ti-6Al-4V interface but the steel|nickel and nickel|copper interfaces were free of intermetallics. During 900 and 925C joining, the copper interlayer could not block diffusion from Ti-6Al-4V to nickel and *vice versa*, so that (Ni,Ti)-based intermetallics formed at the nickel|copper interface within the copper zone. During 950C joining, neither the nickel nor the copper interlayer could block diffusion from Ti-6Al-4V to steel, and *vice versa*. A maximum shear strength of about 377MPa was measured for diffusion couples processed at 875C. The strength of the joints decreased with increasing bonding temperature due to a widening of the distribution of brittle intermetallics in the diffusion zone. Fracture occurred through the Cu_4Ti intermetallic at the copper|Ti-6Al-4V interface when bonded at 875C. After bonding at 900 or 925C, fracture occurred through Ni_3Ti at the nickel|(Ni,Cu,Ti-6Al-4V) reaction interface. After 950C bonding, fracture appeared to occur via σ-phase at the steel|(steel,Ni,Cu,Ti-6Al-4V) reaction interface[422].

X5CrNi18-10|Ti

Grade-2 titanium and the stainless steel were bonded by using 0.1mm-thick nickel foil as the interlayer and heating under vacuum at 850, 875, 900, 925, 950 or 1000C for 1h under a compressive stress of 2MPa. The temperature was the critical factor which controlled the microstructure. The diffusion zone at the borders of the joined materials widened with increasing temperature. The joint structure, progressing from the titanium side, consisted of a titanium plus Ti_2Ni eutectoid and layers of Ti_2Ni, TiNi and $TiNi_3$. The steel|nickel interface was free from any reaction layer at up to 875C while, at higher temperatures, a thin reaction layer appeared[423]. The microhardness across the joint ranged from 320 to 528VHN; values which were higher than those for titanium or stainless steel. Diffusion-bonded (600C, 1h, vacuum, 2, 4, 6 or 8MPa) joints were obtained between titanium and X5CrNi18-10 stainless steel by using 120μm-thick aluminium foil. The intermetallics, $FeAl_3$ and Fe_2Al_5, formed at the stainless-steel|aluminium interfaces, while $TiAl_2$ formed at the aluminium|titanium interfaces. Pressure was an important factor in controlling the mechanical properties: the highest shear strength of 88MPa being found for a bonding pressure of 8MPa[424].

Iron|Tungsten

A further advance is when one of the materials, although metallic, is refractory and not easily melted during joining operations. Tungsten was joined to steel by using a 500μm copper interlayer at 1050C, followed by transient liquid phase diffusion bonding to tungsten using an active 25μm titanium interlayer at 1000C for times of 5, 15, 30, 60 or 180min. With increasing bonding time, the inserted active titanium interlayer, and the Ti-Cu base, tended to be molten and to be consumed by the liquid alloy which formed in the tungsten|steel joint[425]. The tensile strength of the joint meanwhile increased. The strength of the joined specimens could attain 412MPa for a bonding time of 3h.

F82H|W

Tungsten and ferritic steel have been diffusion-bonded using a 0.5mm vanadium interlayer and vacuum hot-pressing (1050C, 1h, 10MPa). The joint had a multilayer sandwich structure which included a tungsten-vanadium transition zone, a residual vanadium interlayer and a diffusion layer between vanadium and the steel. The tungsten-vanadium transition zone consisted mainly of a solid solution. The vanadium/steel diffusion layer had the greatest hardness and comprised a vanadium/VC layer, with a decarburized layer between it and the steel[426]. The tensile strength of the joint attained 75MPa and fractures nucleated mainly in the vanadium/steel interface, because of its brittle VC content. Diffusion bonding (850 to 950C, 1h, vacuum, 10MPa) of tungsten and

F82H ferritic/martensitic steel, with a titanium interlayer, produced excellent joints at both the tungsten|titanium and titanium|F82H interfaces. There was α-β titanium solid solution at the tungsten|titanium interface, and the phases present at the titanium|F82H interface depended upon the joining temperature. The joints fractured at the latter interface during shear testing[427].

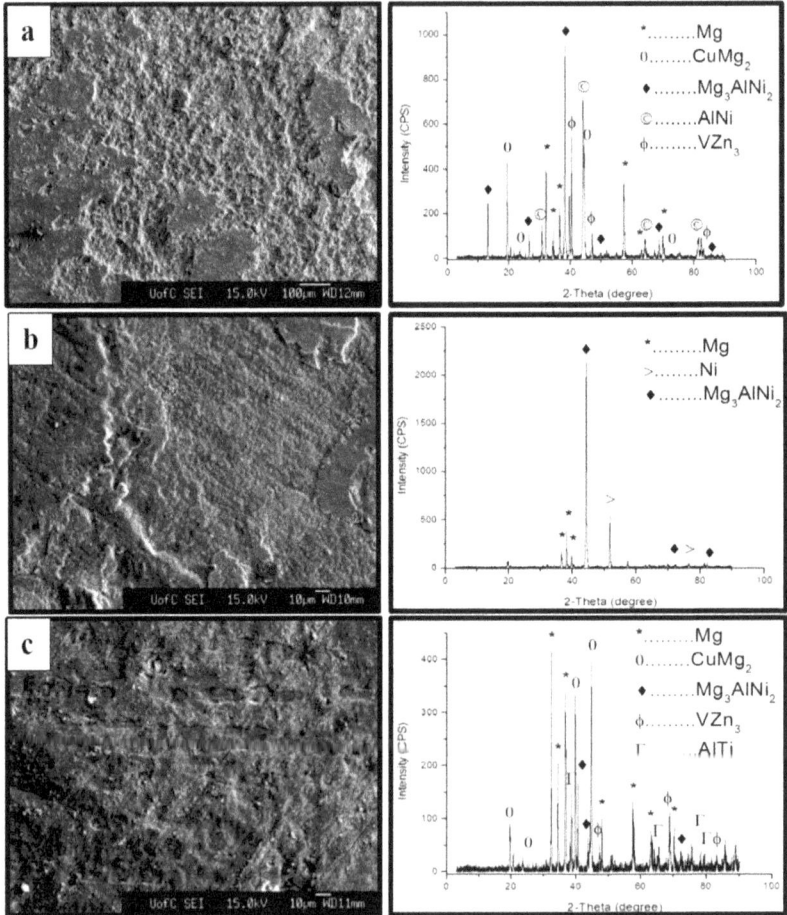

Figure 23. Scanning electron micrographs and X-ray diffraction spectra of fracture surfaces bonded at 515C under 0.35MPa a: Ti-Al-V\NiCu\AZ31 at 10min., b: Ti-Al-V\NiCu\AZ31 at 20min., c: Ti-Al-V\CuNi\AZ31 at 20min.

Transient Liquid Phase Bonding   Materials Research Forum LLC
Materials Research Foundations **43** (2019)   doi: http://dx.doi.org/10.21741/ 9781644900055

AISI316L|W

Transient liquid phase bonding of tungsten and AISI316L stainless steel has been performed by using a Cu-5wt%Ni interlayer at 1120C under vacuum for 600s to 6h, with an applied pressure of 15MPa. The microstructures of joints held for 600s or 0.5h consisted of iron-rich and copper-rich layers with clear boundaries. When the holding time was increased to 3h, the copper-rich layer became thin and dispersed while the iron-rich layer became thick and partially bonded to the stainless-steel austenite grains. When the holding time was 6h, the microstructure and composition of the bonding zone was homogenous and the average shear strength of the joint reached 213MPa; with fracture occurring mainly in the tungsten[428].

Iron|Zirconium

Zircaloy-4 and AISI321 stainless steel have been joined by using an active titanium-based interlayer and heating under vacuum furnace under 1MPa pressure. Control of the heating and cooling rate, and 1200s soaking at 1223K produced a perfect joint. The tensile strength ranged from 480 to 670MPa[429,430]. Simple models were developed in order to predict the evolution of the interlayer during the bonding operation, and the bonds were investigated by means of scanning electron microscopy and energy dispersive X-ray spectrometry. Precision measurements were also made of the interlayer width as a function of bonding temperature. Liquid-film migration occurred due to solubility differences between the stable and metastable phases. The evolution of the interlayer thickness revealed good agreement between the calculations and experimental measurements. Slow isothermal solidification kinetics were not attributed only to enrichment of the liquid phase in alloying elements such as titanium and zirconium, but also to a reduction in the solid solubility limit of copper in the base alloys[431]. The materials were also bonded by using one active titanium-based interlayer and two zirconium-based interlayers under vacuum furnace and 0.5MPa dynamic pressure. Sessile drop tests were performed before bonding at between 820 and 865C for 5, 10 or 15min. A titanium-based interlayer with a higher titanium content exhibited better wetting of the surface of the zirconium alloy at 850C. An increase in the bonding temperature caused an appreciable reduction in the contact angle, spread ratio, spread factor and radius of wetting. The height of the interlayer decreased on the substrate surface upon increasing the bonding temperature. Bonds heated at 850C for 0.25h enjoyed a suitable wetting behaviour but, due to the effect of the bonding temperature, the mechanical properties were impaired and brittle intermetallics formed at the joint interface[432]. The titanium-based interlayer acted as a barrier to the diffusion of iron and chromium and prevented the formation of brittle compounds such as Zr_3Fe_2, Zr_2Ni and Zr_3Fe.

Magnesium|Titanium

AZ31B|Ti-6Al-4V

The magnesium alloy, AZ31B, was first joined to Ti-6Al-4V by using an aluminium interlayer. For a given bonding time of 3h, the reaction products, microstructure and strength depended upon the bonding temperature. When the latter was below 450C, the conditions for Mg-Al eutectic reaction were not satisfied and, since no liquid phase appeared at the joint, bonding was rather unlikely. Temperature had an overall marked effect upon the kinetics of the AZ31B|Al|Ti-6Al-4V interfacial reaction[433]. The joint shear strength of 72.4MPa, after 3h bonding at 470C, amounted to 84.2% of the strength of AZ31B. The transient liquid phase bonding of Ti-6Al-4V to Mg-AZ31 has been performed by using an interlayer which consisted of an electrodeposited nickel coating containing dispersed nickel and copper nanoparticles. The bond formation was attributed to the solid-state diffusion of nickel and magnesium, followed by liquid eutectic formation at the Mg-AZ31 interface. Solid-state diffusion of nickel and titanium at the Ti-6Al-4V interface produced the joint. The use of dispersed copper nanoparticles led to a maximum joint shear strength of 69MPa, corresponding to a 15% enhancement in joint strength over those prepared without the use of a nano-particle dispersion[434]. A similar study was performed using double, nickel and copper, sandwich foils in two configurations: Mg-AZ31|Cu-Ni|Ti-6Al-4V and Mg-AZ31|Ni-Cu|Ti-6Al-4V. Differing reaction layers were formed in the joint region, depending upon the configuration. The formation of ε (Mg), ρ ($CuMg_2$), δ (Mg_2Ni) and Mg_3AlNi_2 phases was found for both configurations. The mechanism of joint formation involved the stages of solid-state diffusion, dissolution and joint-widening, and isothermal solidification[435]. A maximum shear strength of 57MPa was obtained for the Mg-AZ31|Ni-Cu|Ti-6Al-4V configuration. For both configurations, increasing the bonding time decreased the joint strength to 13MPa. Scanning electron micrographs and X-ray diffraction spectra of the fracture surfaces[436] revealed interesting variations in the phases present as a function of the preparation conditions (figure 23).

AZ91D|Ti

The joining of titanium to the magnesium alloy, AZ91D, was studied as a function of bonding time and temperature. Increasing the time from 60s to 1h at 510C caused the copper content of the joint center-line to decrease while the joint microstructure changed from copper solid solution, Cu_2Mg, magnesium solid solution and $CuMg_2$ to magnesium solid solution, $CuMg_2$ and TiC. Increasing the temperature to 530C produced a joint microstructure which consisted of magnesium solid solution, $CuMg_2$ and TiC. Titanium carbide particles aggregated along the joint center-line for bonding times of 0.33 or 1h. An increased fraction of $CuMg_2$, and aggregation of the TiC particles, were the main reasons for a decrease in the joint shear strength[437]. After treatment at 530C for 0.33h, that shear strength reached 69.19MPa; 77.74% of that of the original materials.

Nickel\Titanium

Inconel-738|TiAl

Titanium aluminide was joined to Inconel-718 by using titanium foil, combined with copper or nickel, as an interlayer. Defect-free joints could be prepared, but the liquid formation process was very different for Ti-Cu and Ti-Ni inserts[438]. Sub-layers were found in the joint zone because alloying elements which diffused from the Inconel-718 segregated at the interface between the Inconel and the joint zone. Joints which were produced by using a Ti-Ni insert could solidify isothermally much faster than could those produced by using a Ti-Cu insert.

Ni-|TiAl

Due to the large differences in chemical composition and physical properties, the joining of a titanium aluminide to a nickel-based superalloy is very difficult[439]. Solid-state diffusion bonding (950C, 0.75h, uniaxial pressure: 30MPa) was used to join high niobium-content TiAl alloy. The interfacial microstructure was different to that of the matrix, and changed during diffusion bonding. The interface microhardness was slightly greater than that of the matrix, due to work-hardening of the interface under the uniaxial pressure[440]. Diffusion-bonding tests of high niobium-content TiAl having a fully lamellar microstructure showed that bonding at above 1100C, using a pressure of 30MPa, resulted in recrystallization at the bond interface with a fully lamellar adjacent microstructure; leading to interface migration and sound joints. Recrystallization of the bonded zone improved the quality of the joints by changing the failure mode. The shear strength of the joints reached 400MPa after bonding at 1150C for 0.75h under 30MPa or at 1100C for 0.75h under 40MPa[441]. Diffusion bonding has been used to join a Ti_3Al-based alloy to a nickel-based superalloy, with nickel foil or TiNiNb alloy as an interlayer. In the case of nickel foil, the joint strength first increased and then decreased with increasing bonding temperature. Joints which were formed during 0.33h at 980C, under a pressure of 20MPa, exhibited the maximum room-temperature shear strength of 207MPa. During bonding, the nickel foil reacted with the Ti_3Al-based alloy and the formation of Ni_2Ti, $AlNi_2Ti$ and Ni_3Ti compounds at the Ti_3Al|Ni interface then impaired the mechanical properties of the joint. Use of the TiNiNb alloy interlayer somewhat decreased the incidence of brittle Ti-Ni phases and minimal Ti_2Ni was present in the joint. A shear strength of 209MPa was measured after heating for 600s at 980C under a pressure of 20MPa[442]. The presence of (Ni,Ti,Nb,Fe,Cr) phase at the TiNiNb|GH536 interface was now weakening the joint. Vacuum diffusion bonding (950 to 1100C, 0.33 to 2h, axial pressure: 20MPa) has been used to join Ti_2AlNb and GH4169 by inserting Ni-Nb foil interlayers The Ti_2AlNb and GH4169 were well-bonded, with six intervening reacted layers: Fe-Ni-Cr solid solution | Ni_3Nb | Ni_6Nb_7 | residual Nb | Ti-Nb solid solution | O phase with high niobium content, starting from the GH4169 side. Shear forces caused the join to fracture through the Ni_6Nb_7 layer between the Ni_3Nb layer and the residual niobium layer[443]. The maximum shear strength was 460.3MPa after bonding at 1050C for 0.67h.

NiCr|Ti-6Al-4V

Super-Ni/NiCr laminated composite was joined to Ti-6Al-4V by using vacuum diffusion bonding at 950C. The interface comprised sequential Ni_3Ti, NiTi and $NiTi_2$ layers. The bonding time had a marked effect upon the interfacial microstructure when it was increased from 0.5 to 1.5h. The morphology of the $NiTi_2$ layer changed from serrated to straight and eutectoid Ni_3Ti and $NiTi_2$ formed within the NiTi layer, the width of which significantly increased. The maximum joint shear strength of 69.2MPa was obtained after 1h, and most joints fractured at the super-Ni|Ni_3Ti interface. The shear strength was attributed to plastic deformation and super-Ni crystals[444].

A theoretical analysis of diffusional solidification[445] during the transient liquid phase bonding of dissimilar materials was combined with experiment, leading to a two-dimensional finite element numerical simulation model which did not require the assumption of symmetry. A good agreement with experiment showed that an asymmetrical distribution of residual interlayer liquid during the joining of dissimilar polycrystalline and monocrystalline alloys could be attributed to a mismatch between the lattice diffusion coefficients or solute solubility, regardless of any enhanced intergranular diffusion. In spite of an increase in solute diffusivity with temperature, there could be an increase in bonding temperature and a longer processing period; thus preventing the formation of deleterious eutectic during the bonding of dissimilar materials. The tendency to such anomalous behavior decreased when a material was bonded to one which exhibited a higher solute solubility or a greater ability to accommodate melting-point depressing solute diffusion from the liquid interlayer.

Bonding Ceramics to Ceramics

Partial transient liquid phase bonding can join together hard-to-join materials such as ceramics. The general process uses a multi-layer interlayer comprising a thick refractory core with thin diffusant layers on each side. During heating, the diffusant material melts, and diffusion occurs until the liquid isothermally solidifies. Selecting interlayer materials is a key problem, and tend to be selected *ad hoc*. A selection procedure can suggest a general approach, by linking key characteristics[446]. The selection criteria can be combined with sessile-drop data, leading to the identification of means for increasing bond strength[447].

Ti(C,N)|Al$_2$O$_3$

The bonding of these ceramics was carried out by using Zr$_f$|Cu$_f$|Zr$_f$ sandwiches as an interlayer. The optimum holding time was 0.25 to 0.5h at 950C. At shorter holding times, the joint strength decreased due to the presence of unreacted zirconium at the interface. At longer holding times, the joint strength decreased due to the overgrowth of CuZr intermetallics at the interface. By applying a current pulse during bonding, the residual stress at the interface could be greatly decreased, thus inhibiting the propagation of cracks into the base materials[448]. Such a current pulse could promote the reaction between copper and zinc and the formation of Zr-Cu intermetallics at the interface, producing a weakened joint interface.

Si$_3$N$_4$|SiC

A bond which can withstand up to 415C exposure was formed between SiC and an electroless nickel-plated Si$_3$N$_4$|Cu substrate by using a copper and tin powder paste and heating at 260C. The bond strength at 300C increased with increasing aging time at 300C: from an initial 40MPa to 50MPa after aging for 200h. This increase was attributed to a phase transformation from the Cu$_6$Sn$_5$, formed during joining, to Cu$_3$Sn during aging. An η-(Cu,Ni)$_{6.26}$Sn$_5$ phase was found at the interface between the Cu-Sn bond and a Ni(P) layer[449]. The interface between the Cu-Sn bond and the Si$_3$N$_4$|Cu|Ni(P)|Ag substrate was very stable, with a very slow Ni(P)-layer consumption-rate and Ni$_3$P growth-rate. A Cu-Sn bond on a 5μm-thick electroless plated nickel layer was anticipated to have a lifetime in excess of 105h during storage at 300C.

Bonding Ceramics to Metals

Bonding using a transient liquid phase resembles diffusion welding but requires the use of considerably lower temperatures and pressures. The problem here is to choose an interlayer which will wet the ceramic. When joining zirconia to stainless steel for example, interlayers which comprise zirconium as the active element plus copper or nickel are a potential choice because such systems involve a wide range of constituents with low melting-points which can form the transient liquid phase. Against the above advantages one has to set the facts that the liquid interlayer must wet the ceramic, and that matching the thermal expansion coefficients of the ceramic and metallic interlayers may be difficult. Intermetallic formation is here a *sine qua non*, but thick reaction layers tend to be brittle; thus impairing the strength of the joint. The overall limiting factor is again the wettability of the ceramic.

Numerical simulations of ceramic|metal transient liquid phase bonding have been performed by means of finite element analysis, assuming that an aluminium sheet was used as the interlayer in order to increase wettability[450]. The temperature distribution was predicted by the model, and it was shown that a decrease in the bonding temperature led to a favourable temperature distribution and thus improved the joining of, for example, graphite to copper.

Carbides

C|Nb

The greatest challenge is to join metallic and non-metallic materials. Two-dimensional carbon-carbon composites have been bonded to niobium alloy under vacuum by using a titanium-copper interlayer. The two-stage process involved solid diffusion bonding (780C, 0.5h, 4MPa) and transient liquid-phase diffusion bonding (1050C, 0.5h, 0 or 0.03MPa). The eutectic Ti-Cu liquid which was produced during the second stage exhibited good wettability with respect to the C|C surface, and infiltrated the C|C matrix. A residual copper layer acted as a buffer to relax thermal stress in the joint[451]. The shear strength of the resultant joint could attain 28.6MPa.

C|GH3044

By using Ti/Ni/Cu/Ni multiple foils as an interlayer, a carbon/carbon composite was bonded to the nickel-based superalloy. Multiple interlayers of the form, C-Ti reaction-layer|Ti-Ni intermetallic compound|Ni-Cu|residual copper|Ni-GH3044 diffusion layer, formed between the carbon-carbon composite and the GH3044. The shear strength of the composite|GH3044 joint attained a maximum of 26.1MPa when the bonding temperature used was 1030C[452]. The fracture mode changed with increasing bonding temperature.

C|Ti-6Al-4V

A C_f-SiC composite was bonded to Ti-6Al-4V by heating at 980C for 0.5h and using mixed powders of $Ti_{54.8}Ni_{34.4}Nb_{10.8}$ eutectic and niobium as an interlayer. At the bonding temperature, the latter was a liquid phase with coexisting solid particles. This powder-mixture interlayer was more flexible with regard to joint design, required a shorter bonding time and thus relieved thermal stresses in the joint. The reaction layer consisted of TiC and NbC at the C_f|SiC composite interface, while a diffusion layer appeared at the interface with the Ti-6Al-4V. In the joining layer, residual niobium particles were coated with β-(Ti,Nb) phases and were distributed within a Ti_2Ni matrix together with other massive β-(Ti,Nb) phases. By combining composite-brazing and transient liquid phase bonding into a process termed composite-diffusion brazing, the C_f-SiC composite was bonded (930C, 0.5 to 2h) to Ti-6Al-4V using a (Ti,Zr,Cu,Ni)-16vol%TiC interlayer. Compositional homogenization between the interlayer and the Ti-6Al-4V led to a decrease in the copper and nickel contents in the interlayer, thus markedly improving the heat-resistance of the joint. When the dwell-time was 2h, the melting-point of the joints was increased to 1217C; 300C higher than that of the filler. The shear strength of the joint at 800C attained 137.4MPa for a dwell-time of 1.5h. The combined process was deemed to be better than either of the separate methods.[453,454]

SiC|Cu

Transient liquid phase bonding of silicon carbide to copper was attempted, using metallic lead as an interlayer; that metal being chosen for its high wettability. It was found that the bonding temperature should be 230C in order to obtain good-quality SiC|copper joints[455]. A favorable temperature distribution could be achieved at that temperature and thus the bonding efficiency was optimised.

SiC|Ti-Mo-Nb

Sixteen-ply SiC and Ti-15Mo-2.6Nb-3Al-0.2Si alloy have been bonded by using a 17μm Ti-15Cu-15wt%Ni interlayer and heating, via rapid infra-red heating, at 1100C for 120s, at a ramping rate of 50C/s. Shear-strength testing at up to 800C revealed no joint failure. Due to the inherently rapid cooling of the process, no post-stabilization of the matrix material was required in order to prevent formation of the brittle omega phase during use of the composite at 270 to 430C for up to 20h[456].

TiC|06Cr19Ni10

Titanium carbide cermet was joined to the stainless steel by using a Ti-Cu-Nb interlayer and by employing impulse pressuring to reduce the bonding time. Good joints were obtained at 885C after 2 to 8min under pulsed pressures of 2 to 10MPa. Within the reaction zone, the microstructure exhibited σ-phase, limited solubility of niobium, a sequence of Ti-Cu intermetallic phases, and solid solutions of nickel and copper in α+β titanium. A maximum shear strength of 106.7MPa was obtained after bonding for 300s. Under shear loading, the joints fractured along the Ti-Cu intermetallic interface and spread into the interior of the TiC cermet via brittle cleavage[457].

TiC|AISI304

The vacuum diffusion bonding (925C, 0.33h, 8MPa) of TiC and AISI304 stainless steel, using a Ti/Nb/Cu interlayer, produced a clear transition zone between the carbide and the stainless steel having the sequence: (Ti,Nb) solid solution | Ti | $NbTi_4$ | Nb, residual Cu/(Cu,Fe) solid solution | Cr. The shear strength of the joint ranged up to 84.6MPa, with failure occurring mainly in the TiC|titanium diffusion reaction layer, within the TiC. A niobium interlayer could appreciably relieve residual stresses, so that the interface strength was greater than that of TiC cermets which were weakened by residual stresses[458]. Sintered alumina and AISI304 stainless steel discs were solid-state diffusion-bonded at 900C, using a titanium foil. Precipitation of Ti_3Al particles, plus aluminium and oxygen diffusion, occurred within the titanium; close to the ceramic. Coexisting α-Ti and β-Ti were observed within the titanium foil; being attributed to iron and chromium diffusion, and to β-Ti decomposition during slow cooling from the bonding temperature[459]. A 5μm-thick layer adjacent to the titanium|steel interface contained TiFe, Fe_2Ti, sigma-phase, $M_{23}C_6$ precipitates and TiC.

Ti(C,N)|Ni

These materials were bonded by using a Cu+Nb interlayer under vacuum. The interfacial microstructures initially consisted of Ni_3Nb and a Ni_3Nb plus Cu-Ni solid-solution eutectic. This finally transformed into $(Ti,Nb)(C,N)+Ni_3Nb$ near to the Ti(C,N) and Ni-Cu solid-solution plus Ni_3Nb near to the nickel when the bonding temperature was 1523 to 1573K. Copper clearly acted the part of a constituent in the transient liquid-phase into which nickel dissolved via a Cu-Ni melt. The niobium dissolved rapidly into the Cu-Ni melt. The Ti(C,N) could be wetted by the resultant Ni-Nb-Cu transient liquid phase, with a little (Ti,Nb)(C,N) solid solution being formed at the interface[460]. This increased the interface bonding-ability, and the interface shear strength could attain 140MPa. Theory and experiment showed that the growth of a Ni_3Nb interfacial reaction layer obeyed a parabolic law, that the activation energy for diffusion reaction was 115.0kJ/mol and that the diffusion reaction speed constant was $12.53mm/s^{1/2}$.

WC|40Cr

The cemented carbide, WC-Co, has been bonded to 40Cr steel by using a Ti|Ni|Ti multi-interlayer and heating at 950 to 1100C under vacuum. The two original titanium layers were completely consumed, and transformed into transient liquid phase which reacted with the two materials to be joined. The liquid phase then penetrated into the cemented carbide, leading to the formation of a transition layer of reaction-formed TiC and dispersed WC particles. On the steel side, there was an individual layer of TiC. Layers of

NiTi and Ni_3Ti were found on both sides of the residual nickel core layer. The NiTi layer was also gradually consumed due to continued interfacial reaction as the temperature was increased, together with the growth of TiC and Ni_3Ti layers. The joint shear strength first increased and then sharply decreased when the bonding temperature was higher than 1050C because the Ni_3Ti layer thickened and that was where most joint failures occurred[461]. The maximum strength of 137MPa was obtained by heating at 1000C.

Nitrides

AlN|Aluminium

When producing direct-bonded aluminium substrates using aluminium nitride, a transient eutectic liquid phase formed in the aluminium-X system at the interface between the aluminium foil and the AlN substrate. Here, X was silicon, germanium, silver or copper. The molten aluminium-X phase transiently contacted the AlN substrate before isothermal solidification via diffusion of X into the aluminium foil[462]. The resultant products were very stable following thermal cycling.

AlN|Titanium

Titanium has been joined to aluminium nitride at temperatures as low as 795C by using a commercial brazing alloy. Reactive wetting and spreading occurred at the AlN|braze interface while chemical interaction at the titanium side engendered isothermal solidification of the joint. The governing factor for solidification was the rapid formation of $TiCu_4$ crystals via heterogeneous nucleation and growth in the liquid phase. The brazing alloy was thus depleted in copper, and solid silver precipitated. Following annealing, the re-melting temperature of the joint could be increased up to about 910C; almost 130C higher than the melting point of the original brazing alloy[463].

Si_3N_4|DZ483

The nitride could be joined to the nickel-based DZ483 superalloy by using a Ti|Au|Ni|Ti interlayer. The interfacial microstructure of the joint consisted of three main phases: a TiN reaction phase, a gold-rich phase, and a lamellar nickel-rich phase which was homogenously distributed within the gold-rich phase[464]. A nano-scale waviness of the interface was observed at the Si_3N_4|TiN interface and was thought to explain a high mechanical strength. When a Ni|Cu|Ti interlayer was used, a TiN reaction layer formed at the Si_3N_4|interlayer interface. The reaction layer was composed of two zones: one next to the nitride, with a thickness of about 0.4μm, and another with approximately 0.8μm grains. The microstructure of the joint between the Si_3N_4 and the copper interlayer was described as being a TiN layer with fine grains abutting a TiN layer with coarse grains, in turn abutting a Ti_2Ni layer. Meanwhile Ni_3Ti and $CuTi_2$ were produced at the interface between the copper interlayer and the DZ483 superalloy[465]. The average room-temperature strength of the joint was 147MPa, and a strength of 96MPa was maintained at 1073K. A previous approach, using the Ti|Cu|Ni interlayer, had produced a uniform compact reaction-layer of about 2.7µm in thickness at the Si_3N_4|Ti interface; a result of reaction between titanium and the nitride. With increasing joining temperature, the flexural strength of the joint first increased and then decreased[466]. When the joining temperature was 1323K, the flexural strength of the joint reached 170MPa. All failures occurred in the reaction layer.

$Si_3N_4|Fe$-Al-Cr

The nitride was bonded to an iron aluminide-based alloy (Fe-16Al-5Cr-1Nb-0.05wt%C) by using titanium-enriched 100, 700 or 30µm-thick copper|nickel|aluminium interlayers and heating at 1100 to 1200C under vacuum for between 1.5 and 6h[467]. Following bonding, no low melting-point metal remained at the interface.

$Si_3N_4|Fe$-Cr-Ti

An oxide-dispersion strengthened steel (Fe-20Cr-4.5Al-0.5Ti-0.5wt%Y_2O_3) was bonded to hot-pressed silicon nitride by using an Fe-13B-9wt%Si interlayer at 1200C under vacuum[468]. There was good contact between the liquid interlayer and the ceramic and metal interfaces, and the elemental concentrations around the joint (table 3) were fairly typical of this type of bonding.

Table 3. EDS analysis of joint composition as a function of the distance, x, from the Si_3N_4|interlayer interface

x (µm)	Element	Concentration (wt%)
0	aluminium	8.9
10	aluminium	2.1
20	aluminium	2.6
30	aluminium	2.3
40	aluminium	2.0
50	aluminium	2.4
250	aluminium	2.9

x (µm)	Element	Concentration (wt%)
0	chromium	13.4
10	chromium	16.6
20	chromium	16.7
30	chromium	16.8
40	chromium	26.9
50	chromium	17.2
250	chromium	20.6
0	iron	66.8
10	iron	80.0
20	iron	78.0
30	iron	79.6
40	iron	69.7
50	iron	79.2
250	iron	76.2
0	silicon	7.3
10	silicon	1.1
20	silicon	0.8
30	silicon	0.7
40	silicon	0.8
50	silicon	0.7
250	silicon	0.1
0	titanium	3.6
10	titanium	0.1
20	titanium	0.9
30	titanium	0.3
40	titanium	0.2
50	titanium	0.2
250	titanium	0.2

Si_3N_4|Inconel-718

These materials could be bonded by using nickel as an interlayer. Titanium and copper micro-foils could also be inserted between the nitride and the nickel. The bonding was performed at a lower temperature than normal. Most of the copper precipitated without reacting with the titanium or nickel, while silicon diffused in the interlayer metal and a thin reaction layer formed at the interface between the nitride and the interlayer. This led to good bond-formation and to a high interfacial strength[469]. No interface fracture occurred after cooling from the bonding temperature of 900C. The present bonding method produced more heat-resistant nitride|Inconel-718 joints than did active brazing using Ag-Cu-Ti alloys.

Oxides

The transient liquid phase bonding of an aluminium-based metal-matrix composite and alumina has long been investigated with regard to the connection between particle segregation, copper-interlayer thickness, holding time and joint shear-strength. A long completion-time and slow movement-rate of the solid|liquid interface during bonding greatly increased the tendency to form a segregated-particle layer at the joint interface. Failure occurred via such segregated-particle layers in joints which were produced using 20- or 30µm-thick copper foils and holding times of more than 0.33h. When the segregated-particle layer was less than 10µm, joint failure was governed by the residual stress distribution in the joint and not by preferential fracture through the segregated-particle layer at the bond-line. Good shear strengths were found when a 5µm-thick copper foil was used during transient liquid phase bonding at 853K[470]. Joints between metal-matrix composites, and between such composites and Al_2O_3, were produced at 853K by using 10 to 30µm thick copper foils. The tendency to particle segregation during the bonding of aluminium-based metal-matrix composites greatly increased when the composite contained a large number of small, less than 10µm, reinforcing particles. The particle segregation tendency was much greater when joining Al_2O_3 to metal-matrix composites because the rate of solid|liquid interface movement was much slower and the time required to complete isothermal solidification during transient liquid phase bonding was much longer[471]. The particle-segregation tendency could be avoided by using a short (60s) holding time at 853K, plus subsequent post-weld heat treatment (773K, 4h). This heat-treatment cycle removed any retained eutectic phase from the joint centreline.

Al_2O_3|4J33-Kovar

The 3N-purity alumina was bonded to the 4J33-Kovar by using nickel and titanium foil interlayers. This produced a sandwich microstructure, with α-titanium solid solution in the middle and Ti_2Ni on each side, which imparted good properties to the joint. A Ni_2Ti_4O reaction product was also present at the alumina|interlayer interface, but the structural compatibility of this phase contributed positively to the joint[472]. The highest joint shear strength was about 65MPa, and did not increase monotonically with bonding temperature. Lengthening the bonding time produced a thicker reaction layer and led to a higher joint strength[473].

Al$_2$O$_3$|AISI304

Alumina was first bonded to AlSI304 by using a sandwich filler material which consisted of a tin-based interlayer and amorphous Cu$_{50}$Ti$_{50}$ or Ni-Cr-B interlayers. The highest shear strengths were obtained at a bonding temperature of 500C. Thick defect-containing reaction layers formed at the ceramic|filler interface when higher temperatures were used. Bonding at above 500C also increased the degree of tensile residual stress at the periphery of Al$_2$O$_3$|AlSI304 joints[474]. The shear strength of joints which were produced by using Ni-Cr-B interlayers increased markedly following heat treatment (200C, 1.5h). Such heat treatment had little effect upon the shear strength of joints produced using Cu$_{50}$Ti$_{50}$ interlayers. When joints were made between alumina and the stainless steel by using a nickel or Ni-20wt%Cr interlayer, the residual stresses caused by their thermal-expansion mismatch considerably impaired the joint quality as compared with alumina/alumina joints made under the same conditions[475].

Al$_2$O$_3$|Al-Si

Diffusion bonding (555C, 3h, 50MPa) of Al-5wt%Si and alumina led to joints having average shear strengths which were 25 to 45% of the original strength[476]. When WC-Co cemented carbides were rapidly plasma activated diffusion bonded (750C, 780s, pressure: 40MPa) to 40Cr steels with pure nickel as an interlayer, the original roughness of the joint surfaces had a marked effect upon the microstructure and mechanical behavior. Smoother surfaces largely eliminated interfacial interstices and microvoids. The shear strength of the diffusion-bonded joints thus increased with decreasing surface roughness. The strength decreased sharply when a thicker interlayer was used. A maximum shear strength of 293.07MPa was measured when the original surfaces were ground using 1200-grit SiC paper and a 50μm-thick interlayer was used[477]. Fracture then initiated and propagated mainly along the bonding interfaces instead of within the WC-Co.

Al$_2$O$_3$|Cu

In earlier wetting experiments, molten chromium-nickel-copper powder on Al$_2$O$_3$ substrates was used to investigate the effect of chromium upon the wettability of copper on the Al$_2$O$_3$ substrate. Polycrystalline α-Al$_2$O$_3$ was subsequently joined to AISI304 stainless steel by partial transient liquid phase brazing under vacuum, with chromium or nickel additives in the copper-based filler materials[478]. The results indicated that chromium can improve the properties of the interfaces of ceramic-metal joints. Bonding of Al_2O_3 to copper has been achieved by using a titanium foil interlayer[479]. The interfacial structure was formed by reaction of the Al_2O_3 with Cu-Ti liquid alloy. Aluminium/nickel reactive multilayers with various modulation periods, prepared by direct-current magnetron sputtering, have been used to bond copper to Al_2O_3. The use of such multilayer foils decreased the bonding temperature and improved the joint quality[480]. When bonding (550, 650 or 750C, 0.5h, nitrogen, no pressure) was carried out using those multilayers but with an additional AuSn solder, the combined effect was to reduce the bonding temperature and improve the joint quality[481].

$Al_2O_3|Niobium$

Alumina was bonded to niobium by using titanium and Ni-5V foil interlayers and heating at 1423 to 1573K for 60s to 2h. The shear strength of the joint first increased and then decreased with increasing holding time and bonding temperature. A composite structure, with Nb(V) and Nb(Ti) solid solutions, reinforced by Ni_2Ti, formed when the bonding temperature was 1473K and the holding time was 0.25h; leading to a satisfactory joint strength. Interaction of the titanium and Ni-5V foils led to the formation of a liquid eutectic phase having a low melting point[482]. At the same time, combination of the titanium from the interlayer with oxygen atoms from the Al_2O_3 resulted in the bonding of Al_2O_3 to niobium.

$ZrO_2|Crofer22APU$

Yttrium-stabilised zirconia has been joined to the Fe-Cr alloy by using physical vapor deposited zirconium and titanium as active elements and nickel or copper-nickel foils as core layers[483].

References

[1] Alexandrov, A.S., Quantum, 2[1] 1971, 42-44.

[2] Cao, J., Qi, J., Song, X., Feng, J., Materials, 7[7] 2014, 4930-4962. https://doi.org/10.3390/ma7074930

[3] Paulonis, D.F., Duvall, D.S., Owczarski, W.A., US Patent No.3678570, 25th July 1972

[4] Gale, W.F., Materials Science Forum, 426-432[3] 2003, 1891-1896.

[5] Gale, W.F., Butts, D.A., Science and Technology of Welding and Joining, 9[4] 2004, 283-300. https://doi.org/10.1179/136217104225021724

[6] Shirzadi, A.A., Wallach, E.R., Acta Materialia, 47[13] 1999, 3551-3560. https://doi.org/10.1016/S1359-6454(99)00234-7

[7] Jen, T.C., Jiao, Y., Hwang, T., International Journal of Rotating Machinery, 7[6] 2001, 387-396. https://doi.org/10.1155/S1023621X01000331

[8] Li, J.F., Agyakwa, P.A., Johnson, C.M., Journal of Materials Science, 45[9] 2010, 2340-2350. https://doi.org/10.1007/s10853-009-4199-8

[9] Wang, X., Li, X., Yan, Q., Acta Metallurgica Sinica, 43[10] 2007, 1096-1100.

[10] Illingworth, T.C., Golosnoy, I.O., Clyne, T.W., Materials Science and Engineering A, 445-446, 2007, 493-500. https://doi.org/10.1016/j.msea.2006.09.090

[11] Illingworth, T.C., Golosnoy, I.O., Gergely, V., Clyne, T.W., Journal of Materials Science, 40[9-10] 2005, 2505-2511. https://doi.org/10.1007/s10853-005-1983-y

[12] Sinclair, C.W., Purdy, G.R., Morral, J.E., Metallurgical and Materials Transactions A, 31[4] 2000, 1187-1192. https://doi.org/10.1007/s11661-000-0114-2

[13] Li, X., Wang, X., Yan, F., Wang, H., Wang, X., Applied Mechanics and Materials, 217-219, 2012, 1917-1920.

[14] Natsume, Y., Ohsasa, K., Tayu, Y., Momono, T., Narita, T., ISIJ International, 43[12] 2003, 1976-1982. https://doi.org/10.2355/isijinternational.43.1976

[15] Faiz, M.K., Bansho, K., Suga, T., Miyashita, T., Yoshida, M., Journal of Materials Science - Materials in Electronics, 28[21] 2017, 16433-16443. https://doi.org/10.1007/s10854-017-7554-6

[16] Ghoneim, A., Hunedy, J., Ojo, O.A., Metallurgical and Materials Transactions A, 44[2] 2013, 1139-1151. https://doi.org/10.1007/s11661-012-1412-1

[17] Muhammad, K.F., Yamamoto, T., Yoshida, M., Journal of Materials Science - Materials in Electronics, 28[13] 2017, 9351-9362. https://doi.org/10.1007/s10854-017-6674-3

[18] Faiz, M.K., Yamamoto, T., Yoshida, M., Proceedings of the 5th International Workshop on Low Temperature Bonding for 3D Integration, 2017, 34.

[19] Zhao, H.Y., Liu, J.H., Li, Z.L., Zhao, Y.X., Niu, H.W., Song, X.G., Dong, H.J., Materials Letters, 186, 2017, 283-288. https://doi.org/10.1016/j.matlet.2016.10.017

[20] Liu, J.H., Zhao, H.Y., Li, Z.L., Song, X.G., Dong, H.J., Zhao, Y.X., Feng, J.C., Journal of Alloys and Compounds, 692, 2017, 552-557. https://doi.org/10.1016/j.jallcom.2016.08.263

[21] Bao, Y., Wu, A., Shao, H., Zhao, Y., Zou, G., Journal of Materials Science - Materials in Electronics, 29[12] 2018, 10246-10257. https://doi.org/10.1007/s10854-018-9076-2

[22] Peng, J., Liu, H.S., Ma, H.B., Shi, X.M., Wang, R.C., Journal of Materials Science, 53[12] 2018, 9287-9296. https://doi.org/10.1007/s10853-018-2204-9

[23] Li, J.F., Agyakwa, P.A., Johnson, C.M., Acta Materialia, 59[3] 2011, 1198-1211. https://doi.org/10.1016/j.actamat.2010.10.053

[24] Wei, M., Lin, J., Wu, W., Zhang, D., Heat Treatment of Metals, 37[9] 2012, 79-83.

[25] Bosco, N.S., Zok, F.W., Acta Materialia, 52[10] 2004, 2965-2972. https://doi.org/10.1016/j.actamat.2004.02.043

[26] Bosco, N.S., Zok, F.W., Acta Materialia, 53[7] 2005, 2019-2027. https://doi.org/10.1016/j.actamat.2005.01.013

[27] Park, M.S., Gibbons, S.L., Arróyave, R., Acta Materialia, 60[18] 2012, 6278-6287. https://doi.org/10.1016/j.actamat.2012.07.063

[28] Yoon, J.W., Lee, B.S., Thin Solid Films, 660, 2018, 618-624. https://doi.org/10.1016/j.tsf.2018.04.039

[29] Rahman, A.H.M.E., Cavalli, M.N., Materials Science and Engineering A, 545, 2012, 6-12. https://doi.org/10.1016/j.msea.2012.02.020

[30] Kwon, Y.S., Kim, J.S., Moon, J.S., Suk, M.J., Journal of Materials Science, 35[8] 2000, 1917-1924. https://doi.org/10.1023/A:1004762318057

[31] Lai, Z., Chen, X., Pan, C., Xie, R., Liu, L., Zou, G., Materials Letters, 166, 2016, 219-222. https://doi.org/10.1016/j.matlet.2015.11.031

[32] Xie, R., Chen, X., Lai, Z., Liu, L., Zou, G., Yan, J., Wang, W., Materials and Design, 91, 2016, 19-27. https://doi.org/10.1016/j.matdes.2015.11.071

[33] Jung, J.P., Kang, C.S., Materials Transactions, JIM, 38[10] 1997, 886-891.

[34] Gale, W.F., Wallach, E.R., Metallurgical Transactions A, 22[10] 1991, 2451-2457. https://doi.org/10.1007/BF02665011

[35] Abdelfatah, M., Ojo, O.A., Materials Science and Technology, 25[1] 2009, 61-67. https://doi.org/10.1179/174328407X185929

[36] Saida, K., Zhou, Y., North, T.H., Journal of the Japan Institute of Metals, 58[7] 1994, 810-818. https://doi.org/10.2320/jinstmet1952.58.7_810

[37] Saida, K., Zhou, Y., North, T.H., Journal of Materials Science, 28[23] 1993, 6427-6432. https://doi.org/10.1007/BF01352207

[38] Natsume, Y., Ohsasa, K., Narita, T., Materials Transactions, 44[5] 2003, 819-823. https://doi.org/10.2320/matertrans.44.819

[39] Shinmura, T., Ohsasa, K., Narita, T., Materials Transactions, 42[2] 2001, 292-297. https://doi.org/10.2320/matertrans.42.292

[40] Ikeuchi, K., Zhou, Y., Kokawa, H., North, T.H., Metallurgical Transactions A, 23[10] 1992, 2905-2915. https://doi.org/10.1007/BF02651768

[41] Kokawa, H., Lee, C.H., North, T.H., Metallurgical Transactions A, 22[7] 1991, 1627-1631. https://doi.org/10.1007/BF02667375

[42] Ohsasa, K., Shinmura, T., Narita, T., Journal of Phase Equilibria, 20[3] 1999, 199-206. https://doi.org/10.1361/105497199770335721

[43] Feng, H., Huang, J., Peng, X., Lv, Z., Wang, Y., Yang, J., Chen, S., Zhao, X., Journal of Electronic Materials, 47[8] 2018, 4642-4652. https://doi.org/10.1007/s11664-018-6336-0

[44] Feng, H.L., Huang, J.H., Yang, J., Zhou, S.K., Zhang, R., Wang, Y., Chen, S.H., Electronic Materials Letters, 13[6] 2017, 489-496. https://doi.org/10.1007/s13391-017-6317-0

[45] Feng, H., Huang, J., Yang, J., Zhou, S., Zhang, R., Chen, S., Journal of Electronic Materials, 46[7] 2017, 4152-4159. https://doi.org/10.1007/s11664-017-5357-4

[46] Tuah-Poku, I., Dollar, M., Massalski, T.B., Metallurgical Transactions A, 19[3] 1988, 675-686. https://doi.org/10.1007/BF02649282

[47] Li, J.F., Agyakwa, P.A., Johnson, C.M., Acta Materialia, 58[9] 2010, 3429-3443. https://doi.org/10.1016/j.actamat.2010.02.018

[48] Shao, H., Wu, A., Bao, Y., Zhao, Y., Liu, L., Zou, G., Ultrasonics Sonochemistry, 37, 2017, 561-570. https://doi.org/10.1016/j.ultsonch.2017.02.016

[49] Koyama, S., Takahashi, M., Ikeuchi, K., Materials Science Forum, 539-543[4] 2007, 3503-3507. https://doi.org/10.4028/www.scientific.net/MSF.539-543.3503

[50] Shakerin, S., Maleki, V., Alireza Ziaei, S., Omidvar, H., Rahimipour, M.R., Mirsalehi, S.E., Canadian Metallurgical Quarterly, 56[3] 2017, 360-367. https://doi.org/10.1080/00084433.2017.1349024

[51] Shalz, M.L., Dalgleish, B.J., Tomsia, A.P., Glaeser, A.M., Journal of Materials Science, 29[12] 1994, 3200-3208. https://doi.org/10.1007/BF00356663

[52] Shalz, M.L., Dalgleish, B.J., Tomsia, A.P., Glaeser, A.M., Journal of Materials Science, 28[6] 1993, 1673-1684. https://doi.org/10.1007/BF00363367

[53] Matsumoto, H., Locatelli, M.R., Nakashima, K., Glaeser, A.M., Mori, K., Materials Transactions, JIM, 36[4] 1995, 555-564.

[54] Lo, P.L., Chang, L.S., Lu, Y.F., Ceramics International, 35[8] 2009, 3091-3095. https://doi.org/10.1016/j.ceramint.2009.04.019

[55] Chang, L.S., Huang, C.F., Ceramics International, 30[8] 2004, 2121-2127. https://doi.org/10.1016/j.ceramint.2003.11.018

[56] Sung, M.H., Glaeser, A.M., Proceedings of the 3rd International Brazing and Soldering Conference, 2006, 181-188.

[57] Reynolds, T.B., Bartlow, C.C., Hong, S.M., Glaeser, A.M., TMS Annual Meeting, 2009, 645-652.

[58] Hong, S.M., Bartlow, C.C., Reynolds, T.B., McKeown, J.T., Glaeser, A.M., Advanced Materials, 20[24] 2008, 4799-4803. https://doi.org/10.1002/adma.200801550

[59] Hong, S.M., Reynolds, T.B., Bartlow, C.C., Glaeser, A.M., International Journal of Materials Research, 101[1] 2010, 133-142. https://doi.org/10.3139/146.110249

[60] De Portu, G., Glaeser, A.M., Reynolds, T.B., Takahashi, Y., Boffelli, M., Pezzotti, G., Journal of Materials Science, 50[6] 2014, 2467-2479. https://doi.org/10.1007/s10853-014-8803-1

[61] Dehkordi, O.B., Hadian, A.M., Advanced Materials Research, 829, 2014, 136-140.

[62] Kato, H., Kageyama, K., Materials Science and Technology, 14[7] 1998, 712-718. https://doi.org/10.1179/mst.1998.14.7.712

[63] Wang, J., Li, K., Li, H., Li, W., Li, Z., Guo, L., Journal of Alloys and Compounds, 550, 2013, 57-62. https://doi.org/10.1016/j.jallcom.2012.09.048

[64] Xiong, J.T., Li, J.L., Zhang, F.S., Wang, Z.P., Key Engineering Materials, 336-338[2], 2007, 1260-1262. https://doi.org/10.4028/www.scientific.net/KEM.336-338.1260

[65] Saito, N., Ikeda, H., Yamaoka, Y., Glaeser, A.M., Nakashima, K., Journal of Materials Science, 47[24] 2012, 8454-8463. https://doi.org/10.1007/s10853-012-6778-3

[66] Esposito, L., Sciti, D., Silvestroni, L., Melandri, C., Guicciardi, S., Saito, N., Nakashima, K., Glaeser, A.M., Journal of Materials Science, 49[2] 2014, 654-664. https://doi.org/10.1007/s10853-013-7746-2

[67] Hatami, H.R., Kokabi, A.H., Faghihi Sani, M.A., Proceedings of the 5th International Brazing and Soldering Conference, 2012, 259-265.

[68] Ceccone, G., Nicholas, M.G., Peteves, S.D., Tomsia, A.P., Dalgleish, B.J., Glaeser, A.M., Acta Materialia, 44[2] 1996, 657-667. https://doi.org/10.1016/1359-6454(95)00187-5

[69] Zheng, C., Qizhang, Z., Fang, F., Hongqing, L., Runzhou, S., Zhizhang, L., Transactions of Nonferrous Metals Society of China, 9[4] 1999, 831-837.

[70] Zhou, F., Li, Z., Chinese Journal of Nonferrous Metals, 11[2] 2001, 273-278.

[71] Zou, J.S., Jiang, Z.G., Chu, Y.J., Chen, Z., Journal of Aeronautical Materials, 25[5] 2005, 29-33.

[72] Zou, J., Chu, Y., Xu, Z., Chen, G., China Welding, 13[2] 2004, 100-105.

[73] Zou, J.S., Chu, Y.J., Zhai, J.G., Chen, Z., Transactions of the China Welding Institution, 25[2] 2004, 43-46+51.

[74] Zhou, F., Li, Z., Acta Metallurgica Sinica, 36[2] 2000, 171-176.

[75] Chen, Z., Cao, M.S., Zhao, Q.Z., Zou, J.S., Materials Science and Engineering A, 380[1] 2004, 394-401. https://doi.org/10.1016/j.msea.2004.04.012

[76] Zou, J.S., Chu, Y.J., Xu, Z.R., Chen, Z., Journal of Aeronautical Materials, 24[3] 2004, 43-47.

[77] Zou, J.S., Xu, Z.R., Chu, Y.J., Zhao, Q.Z., Chen, Z., Transactions of the China Welding Institution, 26[2] 2005, 41-44.

[78] Yin, X.H., Li, M.S., Zhou, Y.C., Journal of the European Ceramic Society, 27[12] 2007, 3539-3544. https://doi.org/10.1016/j.jeurceramsoc.2007.01.012

[79] Sciti, D., Silvestroni, L., Esposito, L., Nakashima, K., Saito, N., Yamaoka, Y., Glaeser, A.M., High Temperature Materials and Processes, 31[4-5] 2012, 501-511. https://doi.org/10.1515/htmp-2012-0087

[80] Wei, Y., Sun, F., Tan, S., Liang, S., Vacuum, 154, 2018, 18-24. https://doi.org/10.1016/j.vacuum.2018.04.036

[81] Wang, X.G., Yan, F.J., Li, X.G., Wang, C.G., Science and Technology of Welding and Joining, 22[2] 2017, 170-175. https://doi.org/10.1080/13621718.2016.1209625

[82] Wei, Y., Luo, Y., Qu, H., Tan, S., Zou, J., Liang, S., Special Casting and Nonferrous Alloys, 37[4] 2017, 402-406.

[83] Zhao, J.L., Jie, J.C., Chen, F., Chen, H., Li, T.J., Cao, Z.Q., Transactions of Nonferrous Metals Society of China, 24[6] 2014, 1659-1665. https://doi.org/10.1016/S1003-6326(14)63238-6

[84] Liu, W., Long, L., Ma, Y., Wu, L., Journal of Alloys and Compounds, 643, 2015, 34-39. https://doi.org/10.1016/j.jallcom.2015.04.116

[85] Wang, Y., Luo, G., Li, L., Shen, Q., Zhang, L., Journal of Materials Science, 49[20] 2014, 7298-7308. https://doi.org/10.1007/s10853-014-8440-8

[86] Li, Z., Xu, Z., Zhu, D., Ma, Z., Yan, J., Journal of Materials Processing Technology, 255, 2018, 524-529. https://doi.org/10.1016/j.jmatprotec.2018.01.003

[87] Dong, H.J., Li, Z.L., Song, X.G., Zhao, H.Y., Tian, H., Liu, J.H., Yan, J.C., Materials Science and Engineering A, 705, 2017, 360-365. https://doi.org/10.1016/j.msea.2017.08.099

[88] Mo, L., Zhou, Z., Wu, F., Liu, S., Liu, H., Liu, C., 18th International Conference on Electronic Packaging Technology, 2017, 1046-1050.

[89] Li, Z.L., Dong, H.J., Song, X.G., Zhao, H.Y., Tian, H., Liu, J.H., Feng, J.C., Yan, J.C., Ultrasonics Sonochemistry, 42, 2018, 403-410. https://doi.org/10.1016/j.ultsonch.2017.12.005

[90] Dong, H.J., Li, Z.L., Song, X.G., Zhao, H.Y., Yan, J.C., Tian, H., Liu, J.H., Journal of Alloys and Compounds, 723, 2017, 1026-1031. https://doi.org/10.1016/j.jallcom.2017.06.218

[91] Shao, H., Wu, A., Bao, Y., Transactions of the China Welding Institution, 38[10] 2017, 80-84.

[92] Kuntz, M.L., Zhou, Y., Corbin., S.F., Metallurgical and Materials Transactions A, 37[8] 2006, 2493-2504. https://doi.org/10.1007/BF02586222

[93] Shao, H., Wu, A., Bao, Y., Zhao, Y., Zou, G., Journal of Materials Science - Materials in Electronics, 27[5] 2016, 4839-4848. https://doi.org/10.1007/s10854-016-4366-z

[94] Kuntz, M.L., Panton, B., Wasiur-Rahman, S., Zhou, Y., Corbin, S.F., Metallurgical and Materials Transactions A, 44[8] 2013, 3708-3720. https://doi.org/10.1007/s11661-013-1704-0

[95] Chen, H.Y., Luo, L.M., Zhang, J., Zan, X., Zhu, X.Y., Luo, G.N., Wu, Y.C., Journal of Nuclear Materials, 467, 2015, 566-571. https://doi.org/10.1016/j.jnucmat.2015.10.045

[96] Hu, L., Xue, Y., Shi, F., Materials and Design, 130, 2017, 175-182. https://doi.org/10.1016/j.matdes.2017.05.055

[97] Li, Z., Zhou, Y., North, T.H., Journal of Materials Science, 30[4] 1995, 1075-1082. https://doi.org/10.1007/BF01178448

[98] Li, Z., Fearis, W., North, T.H., Materials Science and Technology, 11[4] 1995, 363-369. https://doi.org/10.1179/mst.1995.11.4.363

[99] Sayyedain, S.S., Salimijazi, H.R., Toroghinejad, M.R., Karimzadeh, F., Materials and Design, 53, 2014, 275-282. https://doi.org/10.1016/j.matdes.2013.06.074

[100] Zhang, G., Zhang, J., Li, S., Wei, Z., Journal of Xian Jiaotong University, 43[11] 2009, 71-74.

[101] Zhang, G., Zhang, J., Pei, Y., Li, S., Chai, D., Materials Science and Engineering A, 488[1-2] 2008, 146-156. https://doi.org/10.1016/j.msea.2007.11.084

[102] Zhang, G., Su, W., Suzumura, A., Metallurgical and Materials Transactions B, 47[3] 2016, 2026-2039. https://doi.org/10.1007/s11663-015-0371-5

[103] Nami, H., Halvaee, A., Adgi, H., Materials and Design, 32[7] 2011, 3957-3965. https://doi.org/10.1016/j.matdes.2011.02.003

[104] Baghbani, M.G., Hadian, A.M., Advanced Materials Research, 829, 2014, 632-637.

[105] Chen, Z., Jin, Z.Y., Gu, X.B., Zou, J.S., Transactions of the China Welding Institution, 22[5] 2001, 27-30.

[106] Chen, R., Zuo, D., Wang, M., Journal of Materials Science and Technology, 22[3] 2006, 291-294.

[107] Chu, Y., Jiang, S., Fan, W., Jin, Z., Wang, D., Advanced Materials Research, 631-

632, 2013, 44-49.

[108] Yarahmadi, M., Sahmanian, M., Salimi Jazi, H.R., Hoshyarmanesh, H.R., Science and Technology of Welding and Joining, 19[7] 2014, 603-608. https://doi.org/10.1179/1362171814Y.0000000231

[109] Wang, B., Jiang, S., Zhang, K., Metallurgical and Materials Transactions A, 43[9] 2012, 3039-3042. https://doi.org/10.1007/s11661-012-1315-1

[110] Askew, J.R., Wilde, J.F., Khan, T.I., Materials Science and Technology, 14[9-10] 1998, 920-924. https://doi.org/10.1179/mst.1998.14.9-10.920

[111] Liu, W.H., Sun, D.Q., Jia, S.S., Qiu, X.M., Transactions of the China Welding Institution, 24[5] 2003, 13-16.

[112] Liu, W., Sun, D., Sun, D., Jia, S., Transactions of the China Welding Institution, 28[3] 2007, 57-60.

[113] Cooke, K.O., Khan, T.I., Oliver, G.D., Metallurgical and Materials Transactions A, 42[8] 2011, 2271-2277. https://doi.org/10.1007/s11661-011-0663-6

[114] Cooke, K.O., Khan, T.I., Oliver, G.D., Science and Technology of Welding and Joining, 17[1] 2012, 22-31. https://doi.org/10.1179/1362171811Y.0000000069

[115] Cooke, K.O., Khan, T.I., Oliver, G.D., Materials and Design, 33[1] 2012, 469-475. https://doi.org/10.1016/j.matdes.2011.04.051

[116] Cooke, K.O., Metallurgical and Materials Transactions B, 43[3] 2012, 627-634. https://doi.org/10.1007/s11663-012-9643-5

[117] Sabathier, V., Edwards, G.R., Cross, C.E., Metallurgical and Materials Transactions A, 25[12] 1994, 2705-2714. https://doi.org/10.1007/BF02649223

[118] Liu, L.M., Gao, Z.K., Chinese Journal of Nonferrous Metals, 15[6] 2005, 849-853.

[119] Maity, J., Pal, T.K., Maiti, R., Journal of Materials Processing Technology, 209[7] 2009, 3568-3580. https://doi.org/10.1016/j.jmatprotec.2008.08.015

[120] Maity, J., Pal, T.K., Maiti, R., Materials Science and Technology, 25[12] 2009, 1489-1494. https://doi.org/10.1179/174328409X407551

[121] Maity, J., Pal, T.K., Maiti, R., Journal of Materials Science, 45[13] 2010, 3575-3587. https://doi.org/10.1007/s10853-010-4402-y

[122] Barman, S.C., Pal, T.K., Maity, J., Materials Science and Technology, 27[5] 2011, 951-957. https://doi.org/10.1179/026708310X12712410311893

[123] Maity, J., Pal, T.K., Journal of Materials Engineering and Performance, 21[7] 2012, 1232-1242. https://doi.org/10.1007/s11665-011-0037-7

[124] Roy, P., Pal, T.K., Maity, J., Journal of Materials Engineering and Performance, 25[8] 2016, 3518-3530. https://doi.org/10.1007/s11665-016-2165-6

[125] Shirzadi, A.A., Wallach, E.R., Materials Science and Technology, 13[2] 1997, 135-142. https://doi.org/10.1179/mst.1997.13.2.135

[126] Hua, M., Guo, W., Law, H.W., Ho, J.K.L., International Journal of Advanced Manufacturing Technology, 37[5-6] 2008, 504-512. https://doi.org/10.1007/s00170-007-0983-2

[127] Zhang, G., Cai, J., Chen, B., Xu, T., Materials and Design, 110, 2016, 653-662. https://doi.org/10.1016/j.matdes.2016.08.024

[128] Zhang, G., Chen, B., Jin, M., Zhang, J., Materials Transactions, 56[2] 2015, 212-217. https://doi.org/10.2320/matertrans.M2014295

[129] Huang, Z., Du, H., Liu, L., Lai, Z., Chen, X., Long, W., Wang, W., Zou, G., Ultrasonics Sonochemistry, 43, 2018, 101-109. https://doi.org/10.1016/j.ultsonch.2017.11.022

[130] Gu, X.Y., Sun, D.Q., Liu, L., Materials Science and Engineering A, 487[1-2] 2008, 86-92. https://doi.org/10.1016/j.msea.2007.09.064

[131] Gu, X., Sun, D., Liu, L., China Welding, 16[1] 2007, 19-24.

[132] Zhang, L., Hou, J.B., Journal of Aeronautical Materials, 26[3] 2006, 325-326.

[133] Fukumoto, S., Imamura, K., Hirose, A., Kobayashi, K.F., ISIJ International, 35[10] 1985, 1307-1314. https://doi.org/10.2355/isijinternational.35.1307

[134] Fukumoto, S., Kasahara, A., Hirose, A., Kobayashi, K.F., Materials Science and Technology, 10[9] 1994, 807-812. https://doi.org/10.1179/mst.1994.10.9.807

[135] Wang, X.G., Li, X.G., Wang, C.G., Science and Technology of Welding and Joining, 17[5] 2012, 414-418. https://doi.org/10.1179/1362171812Y.0000000026

[136] Jabbareh, M.A., Assadi, H., Scripta Materialia, 60[9] 2009, 780-782. https://doi.org/10.1016/j.scriptamat.2009.01.013

[137] Sekerka, R.F., International Association of Hydrological Sciences, 1980, 1-3.

[138] Eluri, R., Paul, B.K., Materials and Design, 36, 2012, 13-23. https://doi.org/10.1016/j.matdes.2011.11.005

[139] Schällibaum, J., Burbach, T., Münch, C., Weiler, W., Wahlen, A., Materialwissenschaft und Werkstofftechnik, 46[7] 2015, 704-712. https://doi.org/10.1002/mawe.201500402

[140] Assadi, H., Shirzadi, A.A., Wallach, E.R., Acta Materialia, 49[1] 2001, 31-39. https://doi.org/10.1016/S1359-6454(00)00307-4

[141] Afghahi, S.S.S., Ekrami, A., Farahany, S., Jahangiri, A., Transactions of Nonferrous Metals Society of China, 25[4] 2015, 1073-1079. https://doi.org/10.1016/S1003-6326(15)63700-1

[142] Yang, C.F., Chiu, L.H., Lee, S.C., Sun, J.Y., Proceedings of the National Science Council, Republic of China - Physical Science and Engineering, 22[1] 1998, 132-141.

[143] Dunford, D.V., Partridge, P.G., Materials Science and Technology, 14[5] 1998, 422-428. https://doi.org/10.1179/mst.1998.14.5.422

[144] Jen, T.C., Jiao, Y., Proceedings of the National Heat Transfer Conference, 1, 2001, 475-481.

[145] Jen, T.C., Jiao, Y., Numerical Heat Transfer A, 39[2] 2001, 123-138. https://doi.org/10.1080/104077801300004230

[146] Zhang, W., Qiu, X., Chen, X., Zhao, X., Sun, D., Li, Y., Transactions of the China Welding Institution, 30[2] 2009, 121-124.

[147] Sun, Q., Wang, H., Yang, C., Metallurgical and Materials Transactions B, 49[3] 2018, 933-938. https://doi.org/10.1007/s11663-018-1264-1

[148] Iskandar, R., Schwedt, A., Mayer, J., Rochala, P., Wiesner, S., Öte, M., Bobzin, K., Weirich, T.E., Materialwissenschaft und Werkstofftechnik, 48[12], 2017, 1257-1263. https://doi.org/10.1002/mawe.201700155

[149] Wang, Q., Chen, X., Zhu, L., Yan, J., Lai, Z., Zhao, P., Bao, J., Lv, G., You, C., Zhou, X., Zhang, J., Li, Y., Ultrasonics Sonochemistry, 34, 2017, 947-952. https://doi.org/10.1016/j.ultsonch.2016.08.004

[150] Atabaki, M.M., Journal of Materials Engineering and Performance, 21[6] 2012, 1040-1045.

[151] Atabaki, M.M., Welding Journal, 91[2] 2012, 35s-49s.

[152] Venkateswaran, T., Ravi, K.R., Sivakumar, D., Pant, B., Janaki Ram, G.D., Journal of Materials Engineering and Performance, 26[8] 2017, 4064-4071. https://doi.org/10.1007/s11665-017-2830-4

[153] Di Luozzo, N., Boudard, M., Fontana, M., Arcondo, B., Materials and Design, 92, 2016, 760-766. https://doi.org/10.1016/j.matdes.2015.12.101

[154] Di Luozzo, N., Martínez Stenger, P.F., Canal, J.P., Fontana, M.R., Arcondo, B., Hyperfine Interactions, 203[1-3] 2011, 125-132. https://doi.org/10.1007/s10751-011-0372-y

[155] Di Luozzo, N., Boudard, M., Doisneau, B., Fontana, M., Arcondo, B., Journal of Alloys and Compounds, 615[S1] 2014, S13-S17. https://doi.org/10.1016/j.jallcom.2013.11.165

[156] Di Luozzo, N., Doisneau, B., Boudard, M., Fontana, M., Arcondo, B., Journal of Alloys and Compounds, 615[S1], 2014, S18-S22. https://doi.org/10.1016/j.jallcom.2013.11.161

[157] Di Luozzo, N., Fontana, M., Arcondo, B., Journal of Materials Science, 42[11] 2007, 4044-4050. https://doi.org/10.1007/s10853-006-0190-9

[158] Di Luozzo, N., Fontana, M., Arcondo, B., Journal of Materials Science, 43[14] 2008, 4938-4944. https://doi.org/10.1007/s10853-008-2720-0

[159] Li, H., Li, Z.X., Journal of Sandwich Structures and Materials, 10[3] 2008, 247-266. https://doi.org/10.1177/1099636208089309

[160] Li, H., Li, Z.X., Journal of Materials Engineering and Performance, 17[6] 2008, 849-856. https://doi.org/10.1007/s11665-008-9244-2

[161] Husain, M.M., Ghosh, M., International Journal of Advanced Manufacturing Technology, 66[9-12] 2013, 1871-1877. https://doi.org/10.1007/s00170-012-4466-8

[162] Tang, H., Chen, S., Transactions of the China Welding Institution, 33[3] 2012, 101-104.

[163] Wang, X.G., Yan, Q., Li, X.G., Transactions of the China Welding Institution, 26[5] 2005, 56-60.

[164] Noto, H., Kasada, R., Kimura, A., Ukai, S., Journal of Nuclear Materials, 442[1-3, S1] 2013, S567-S571.

[165] Noto, H., Ukai, S., Hayashi, S., Journal of Nuclear Materials, 417[1-3] 2011, 249-252. https://doi.org/10.1016/j.jnucmat.2010.12.057

[166] Noh, S., Kasada, R., Kimura, A., Green Energy and Technology, 66, 2011, 292-299. https://doi.org/10.1007/978-4-431-53910-0_39

[167] Noh, S., Kasada, R., Kimura, A., Green Energy and Technology, 44, 2010, 339-345. https://doi.org/10.1007/978-4-431-99779-5_55

[168] Wang, X., Yan, Q., Li, X., Transactions of the China Welding Institution, 28[5] 2007, 53-56.

[169] Wang, X.G., Li, X.G., Yan, Q., Science and Technology of Welding and Joining, 12[5] 2007, 455-459. https://doi.org/10.1179/174329307X213891

[170] Wang, X., Yan, Q., Li, X., IET Conference Publications, 524, 2006, 32-34.

[171] Rhee, B.H., Kim, D.U., Metals and Materials International, 8[5] 2002, 427-433. https://doi.org/10.1007/BF03027238

[172] Chen, S.J., Jing, X.T., Li, X.G., Transactions of the China Welding Institution, 25[6] 2004, 73-76.

[173] Chen, S.J., Jing, X.T., Li, X.G., Transactions of the China Welding Institution, 26[4] 2005, 69-72.

[174] Chen, S.J., Tang, H.J., Jing, X.T., Materials Science and Engineering A, 499[1-2] 2009, 114-117. https://doi.org/10.1016/j.msea.2007.11.133

[175] AlHazaa, A.N., Shar, M.A., Atieh, A.M., Nishikawa, H., Metals, 8[1] 2018, 60-69. Image reproduced under licence https://creativecommons.org/licenses/by/4.0/

[176] AlHazaa, A.N., Khalil, K.A., Shar, M.A., Journal of King Saud University - Science, 28[2] 2016, 152-159.

[177] Jin, Y.J., Khan, T.I., Materials and Design, 38, 2012, 32-37. https://doi.org/10.1016/j.matdes.2012.01.039

[178] Sun, D.Q., Gu, X.Y., Liu, W.H., Materials Science and Engineering A, 391[1-2] 2005, 29-33. https://doi.org/10.1016/j.msea.2004.06.008

[179] Sun, D.Q., Liu, W.H., Gu, X.Y., Materials Science and Technology, 20[12] 2004, 1595-1598. https://doi.org/10.1179/174328413X13789824293506

[180] Lai, Z., Xie, R., Pan, C., Chen, X., Liu, L., Wang, W., Zou, G., Journal of Materials Science and Technology, 33[6] 2017, 567-572. https://doi.org/10.1016/j.jmst.2016.11.002

[181] Campbell, C.E., Boettinger, W.J., Metallurgical and Materials Transactions A, 31[11] 2000, 2835-2847. https://doi.org/10.1007/BF02830355

[182] Gale, W.F., Orel, S.V., Journal of Materials Science, 31[2] 1996, 345-349. https://doi.org/10.1007/BF01139150

[183] Gale, W.F., Guan, Y., Metallurgical and Materials Transactions A, 27[11] 1996, 3621-3629. https://doi.org/10.1007/BF02595453

[184] Payton, L.N., Aluru, R., Gale, W.F., Sofyan, N.I., Krishnardula, V.G., Butts, D.A., Chitti, S.V., Zhou, T., ASM Conference Proceedings: Joining of Advanced and Specialty Materials, 2004, 85-90.

[185] Sheng, N., Hu, X., Liu, J., Jin, T., Sun, X., Hu, Z., Metallurgical and Materials Transactions A, 46[4] 2015, 1670-1677. https://doi.org/10.1007/s11661-014-2733-z

[186] Sheng, N., Liu, J., Jin, T., Sun, X., Hu, Z., Philosophical Magazine, 94[11] 2014, 1219-1234. https://doi.org/10.1080/14786435.2014.885136

[187] Sheng, N., Liu, J., Jin, T., Sun, X., Hu, Z., Metallurgical and Materials Transactions A, 46[12] 2015, 5772-5781. https://doi.org/10.1007/s11661-015-3154-3

[188] Pouranvari, M., Materials Science and Technology, 31[14] 2015, 1773-1780. https://doi.org/10.1179/1743284715Y.0000000005

[189] Duan, H., Luo, J., Bohm, K.H., Kocak, M., Journal of University of Science and Technology Beijing - Mineral Metallurgy Materials, 12[5] 2005, 431-435.

[190] Duan, H., Koçak, M., Bohm, K.H., Ventzke, V., Science and Technology of Welding and Joining, 9[6] 2004, 513-518. https://doi.org/10.1179/136217104225021850

[191] Lin, T., Li, H., He, P., Wei, H., Li, L., Feng, J., Intermetallics, 37, 2013, 59-64. https://doi.org/10.1016/j.intermet.2013.01.015

[192] Peng, W., Zhang, D., Tang, D., Lin, J., Heat Treatment of Metals, 39[8] 2014, 5-10.

[193] Butts, D.A., Gale, W.F., Materials Science and Technology, 24[12] 2008, 1492-1500. https://doi.org/10.1179/174328407X236535

[194] Butts, D.A., Gale, W.F., ASM Proceedings of the International Conference: Trends in Welding Research, 2005, 867-872.

[195] Gu, X., Sun, D., Ren, Z., Liu, L., Duan, Z., Transactions of the China Welding Institution, 31[5] 2010, 45-48.

[196] Gu, X.Y., Sun, D.Q., Liu, L., Duan, Z.Z., Material Science and Technology, 17[S1] 2009, 138-142.

[197] Gale, W.F., Wen, X., Zhou, T., Shen, Y., Materials Science and Technology, 17[11] 2001, 1423-1433. https://doi.org/10.1179/026708301101509403

[198] Butts, D.A., Gale, W.F., Zhou, T., ASM Proceedings of the International Conference: Trends in Welding Research, 2002, 495-499.

[199] Fergus, J.W., Zhou, T., Dang, B., Gale, W.F., ASM Proceedings of the International Conference: Trends in Welding Research, 2002, 799-803.

[200] Gale, W.F., Butts, D.A., Di Ruscio, M., Zhou, T., Metallurgical and Materials Transactions A, 33[10] 2002, 3205-3214. https://doi.org/10.1007/s11661-002-0306-z

[201] Zhou, T., Gale, W.F., Butts, D., Di Ruscio, M., ASM Conference Proceedings: Joining of Advanced and Specialty Materials, 2001, 51-55.

[202] Cai, X., Wang, Y., Yang, Z., Wang, D., Transactions of the China Welding Institution, 39[2] 2018, 24-28.

[203] Cai, X.Q., Wang, Y., Yang, Z.W., Wang, D.P., Liu, Y.C., Journal of Alloys and Compounds, 679, 2016, 9-17. https://doi.org/10.1016/j.jallcom.2016.03.286

[204] Shen, L.J., Liu, F.C., Yang, G.L., Huang, Y.D., Ke, L.M., Materials Science Forum, 898, 2017, 1247-1253. https://doi.org/10.4028/www.scientific.net/MSF.898.1247

[205] Wang, G., Huang, Y., Wang, G., Shen, J., Chen, Z., Journal of Wuhan University of Technology - Materials Science, 30[3] 2015, 617-621.

[206] Deng, Y., Sheng, G., Wang, F., Yuan, X., An, Q., Materials and Design, 92, 2016, 1-7. https://doi.org/10.1016/j.matdes.2015.11.103

[207] Zou, G.S., Xie, E.H., Bai, H.L., Wu, A.P., Wang, Q., Ren, J.L., Materials Science and Engineering A, 499[1-2] 2009, 101-105. https://doi.org/10.1016/j.msea.2007.11.104

[208] Kejanli, H., Taşkin, M., Kolukisa, S., Topuz, P., International Journal of Advanced Manufacturing Technology, 44[7-8] 2009, 695-699. https://doi.org/10.1007/s00170-008-1860-3

[209] Bakhtiari, R., Ekrami, A., Materials Characterization, 66, 2012, 38-45. https://doi.org/10.1016/j.matchar.2012.02.002

[210] Bakhtiari, R., Ekrami, A., Materials and Design, 40, 2012, 130-137. https://doi.org/10.1016/j.matdes.2012.03.036

[211] Zhang, L., Hou, J., Zhang, S., China Welding, 16[1] 2007, 63-67.

[212] Khan, T.I., Orhan, N., Eroglu, M., Materials Science and Technology, 18[4] 2002, 396-400. https://doi.org/10.1179/026708302225001697

[213] Khan, T.I., Kabir, M.J., Bulpett, R., Materials Science and Engineering A, 372[1-2] 2004, 290-295. https://doi.org/10.1016/j.msea.2004.01.023

[214] Padron, T., Khan, T.I., Kabir, M.J., Materials Science and Engineering A, 385[1-2] 2004, 220-228. https://doi.org/10.1016/S0921-5093(04)00860-3

[215] Suzumura, A., Onzawa, T., Tamura, H., Journal of the Japan Welding Society, 50[7] 1981, 646-652. https://doi.org/10.2207/qjjws1943.50.646

[216] Atabaki, M.M., Wati, J.N., Idris, J., Welding Journal, 92[3] 2013, 57s-63s.

[217] Atabaki, M.M., Wati, J.N., Idris, J.B., ASM Heat Treating Society - 26th Conference and Exposition, 2011, 20-43.

[218] Hdz-García, H.M., Martinez, A.I., Mu-oz-Arroyo, R., Acevedo-Dávila, J.L., García-Vázquez, F., Reyes-Valdes, F.A., Journal of Materials Science and Technology, 30[3] 2014, 259-262. https://doi.org/10.1016/j.jmst.2013.11.006

[219] Sadeghian, M., Ekrami, A., Jamshidi, R., Science and Technology of Welding and Joining, 22[8] 2017, 666-672. https://doi.org/10.1080/13621718.2017.1302180

[220] Zhuang, W.D., Eagar, T.W., Welding Journal, 76[12] 1997, 157.

[221] Zhou, X., Dong, Y., Liu, C., Liu, Y., Yu, L., Chen, J., Li, H., Yang, J., Materials and Design, 88, 2015, 1321-1325. https://doi.org/10.1016/j.matdes.2015.09.104

[222] Wang, X.G., Yan, F.J., Yan, Q., Li, X.G., Journal of Iron and Steel Research, 20[4] 2008, 43-46.

[223] Rhee, B., Roh, S., Kim, D., Materials Transactions, 44[5] 2003, 1014-1023. https://doi.org/10.2320/matertrans.44.1014

[224] Yuan, X.J., Kang, C.Y., Materials Research Innovations, 17[S1], 2013, 297-300. https://doi.org/10.1179/1432891713Z.000000000234

[225] Yuan, X., Kim, M.B., Kang, C.Y., Materials Characterization, 60[11] 2009, 1289-1297. https://doi.org/10.1016/j.matchar.2009.05.012

[226] Yuan, X., Kim, M.B., Kang, C.Y., Metallurgical and Materials Transactions A, 42[5] 2011, 1310-1324. https://doi.org/10.1007/s11661-010-0534-6

[227] Yuan, X.J., Luo, J., Li, J., Kang, C.Y., Journal of Central South University - Science and Technology, 43[10] 2012, 3814-3819.

[228] Khan, T.I., Wallach, E.R., Journal of Materials Science, 30[20] 1995, 5151-5160. https://doi.org/10.1007/BF00356063

[229] Nakao, Y., Shinozaki, K., Materials Science and Technology, 11[3] 1995, 304-311. https://doi.org/10.1179/mst.1995.11.3.304

[230] Khan, T.I., Wallach, E.R., Materials Science and Technology, 12[7] 1996, 603-606. https://doi.org/10.1179/mst.1996.12.7.603

[231] Khan, T.I., Wallach, E.R., Journal of Materials Science, 31[11] 1996, 2937-2943. https://doi.org/10.1007/BF00356005

[232] Krishnardula, V.G., Sofyan, N.I., Fergus, J.W., Gale, W.F., ASM Proceedings of the International Conference: Trends in Welding Research, 2005, 861-865.

[233] Krishnardula, V.G., Sofyan, N.I., Gale, W.F., Fergus, J.W., Metallurgical and Materials Transactions A, 37[2] 2006, 497-500. https://doi.org/10.1007/s11661-006-0021-2

[234] Zhang, S., Hou, J.B., Guo, D.L., Zhang, L., Transactions of the China Welding Institution, 25[3] 2004, 43-47.

[235] Wu, S., Hou, J.B., Teng, J.F., Yang, Y., Material Science and Technology, 17[S1] 2009, 33-35.

[236] Nishimoto, K., Saida, K., Kim, D., Nakao, Y., ISIJ International, 35[10] 1995, 1298-1306. https://doi.org/10.2355/isijinternational.35.1298

[237] Nishimoto, K., Kim, D., Saida, K., Asai, S., Furukawa, Y., Nakao, Y., Quarterly Journal of the Japan Welding Society, 14[4] 1996, 731-740. https://doi.org/10.2207/qjjws.14.731

[238] Nishimoto, K., Saida, K., Kim, D., Asai, S., Furukawa, Y., Quarterly Journal of the Japan Welding Society, 15[3] 1997, 515-522. https://doi.org/10.2207/qjjws.15.515

[239] Nishimoto, K., Saida, K., Kim, D., Asai, S., Furukawa, Y., Quarterly Journal of the Japan Welding Society, 15[3] 1997, 509-514. https://doi.org/10.2207/qjjws.15.509

[240] Nishimoto, K., Saida, K., Kim, D., Asai, S., Furukawa, Y., Quarterly Journal of the Japan Welding Society, 15[2] 1997, 321-329. https://doi.org/10.2207/qjjws.15.321

[241] Nishimoto, K., Saida, K., Kim, D., Asai, S., Furukawa, Y., Quarterly Journal of the Japan Welding Society, 16[4] 1998, 530-539. https://doi.org/10.2207/qjjws.16.530

[242] Nishimoto, K., Saida, K., Kim, D., Asai, S., Furukawa, Y., Nakao, Y., Welding in the World, 41[2] 1998, 121-131.

[243] Nishimoto, K., Saida, K., Kim, D., Asai, S., Furukawa, Y., Nakao, Y., Welding Research Abroad, 44[11-12] 1998, 48-58.

[244] Nishimoto, K., Saida, K., Kim, D., Asai, S., Furukawa, Y., Quarterly Journal of the Japan Welding Society, 18[1] 2000, 133-140. https://doi.org/10.2207/qjjws.18.133

[245] Kim, D.U., Nishimoto, K., Metals and Materials International, 8[4] 2002, 403-410. https://doi.org/10.1007/BF03186114

[246] Kim, D.U., Nishimoto, K., Materials Science and Technology, 19[4] 2003, 456-460. https://doi.org/10.1179/026708303225001957

[247] Kim, D.S., Lee, H.S., Yoo, K.B., Song, K.S., Materials Science and Technology Conference and Exhibition 2011, 1507-1510.

[248] Yu, Z., Ding, X., Cao, L., Zheng, Y., Feng, Q., Acta Metallurgica Sinica, 52[5] 2016, 549-560.

[249] Fujita, Y., Saida, K., Kusano, M., Nishimoto, K., Quarterly Journal of the Japan Welding Society, 24[3] 2006, 233-239. https://doi.org/10.2207/qjjws.24.233

[250] Fujita, Y., Saida, K., Kusano, M., Nishimoto, K., Quarterly Journal of the Japan Welding Society, 25[1] 2007, 31-37. https://doi.org/10.2207/qjjws.25.31

[251] Li, X.H., Zhong, Q.P., Cao, C.X., Journal of Aeronautical Materials, 23[2] 2003, 1-5+24.

[252] Li, X.H., Mao, W., Guo, W.L., Xie, Y.H., Transactions of the China Welding Institution, 26[4] 2005, 51-54.

[253] Li, X., Mao, W., Guo, W., Xie, Y., Ye, L., Cheng, Y., China Welding, 14[1] 2005, 19-23.

[254] Nishimoto, K., Saida, K., Shinohara, Y., Science and Technology of Welding and Joining, 8[1] 2003, 29-38. https://doi.org/10.1179/136217103225008946

[255] Lang, B., Hou, J.B., Wu, S., Journal of Materials Engineering, 10, 2010, 32-37.

[256] Li, X.H., Mao, W., Cheng, Y.Y., Ma, W.L., Welding in the World, 49[1-2] 2005, 34-38. https://doi.org/10.1007/BF03266462

[257] Pan, L., Xing, L., Liu, F., Yang, C., Transactions of the China Welding Institution, 34[4] 2013, 89-92.

[258] Choi, W., Kim, S., Lee, C., Jang, J., Materials Science Forum, 449-452[1] 2004, 133-136. https://doi.org/10.4028/www.scientific.net/MSF.449-452.133

[259] Lee, B.K., Song, W.Y., Kim, D.U., Woo, I.S., Kang, C.Y., Metals and Materials International, 13[1] 2007, 59-65. https://doi.org/10.1007/BF03027824

[260] Pouranvari, M., Ekrami, A., Kokabi, A.H., Journal of Alloys and Compounds, 461[1-2] 2008, 641-647. https://doi.org/10.1016/j.jallcom.2007.07.108

[261] Pouranvari, M., Materiali in Tehnologije, 48[1] 2014, 113-118.

[262] Lee, B.K., Oh, I.S., Kim, G.M., Kang, C.Y., Journal of Korean Institute of Metals and Materials, 47[4], 2009, 242-247.

[263] Kim, J.K., Park, H.J., Shim, D.N., Kim, D.J., Journal of Manufacturing Processes, 25, 2017, 60-69. https://doi.org/10.1016/j.jmapro.2016.10.002

[264] Kapoor, M., Doğan, Ö.N., Carney, C.S., Saranam, R.V., McNeff, P., Paul, B.K., Metallurgical and Materials Transactions A, 48[7] 2017, 3343-3356. https://doi.org/10.1007/s11661-017-4127-5

[265] Ghoneim, A., Ojo, O.A., 7th International Symposium on Superalloy 718 and Derivatives 2010, 427-438.

[266] Ghoneim, A., Ojo, O.A., Materials Characterization, 62[1] 2011, 1-7. https://doi.org/10.1016/j.matchar.2010.09.011

[267] Li, X.H., Mao, W., Cheng, Y.Y., Transactions of Nonferrous Metals Society of China, 11[3] 2001, 405-408.

[268] Wu, S., Hou, J., Lang, B., Transactions of the China Welding Institution, 33[2] 2012, 105-108.

[269] Lang, B., Hou, J., Wu, S., Transactions of the China Welding Institution, 33[8] 2012, 109-112.

[270] Jalilian, F., Jahazi, M., Drew, R.A.L., Materials Science and Engineering A, 423[1-2], 2006, 269-281. https://doi.org/10.1016/j.msea.2006.02.030

[271] Jalilian, F., Jahazi, M., Drew, R.A.L., Metallography, Microstructure, and Analysis, 2[3] 2013, 170-182. https://doi.org/10.1007/s13632-013-0070-z

[272] Esmaeili, H., Mirsalehi, S.E., Farzadi, A., Metallurgical and Materials Transactions B, 48[6] 2017, 3259-3269. https://doi.org/10.1007/s11663-017-1098-2

[273] Murray, D.C., Corbin, S.F., Journal of Materials Processing Technology, 248, 2017, 92-102. https://doi.org/10.1016/j.jmatprotec.2017.05.013

[274] Yeh, M.S., Chuang, T.H., Welding Journal, 76[12] 1997, 517.

[275] Sakamoto, R., Nakanishi, T., Saida, K., Nishimoto, K., Quarterly Journal of the Japan Welding Society, 24[3] 2006, 273-280. https://doi.org/10.2207/qjjws.24.273

[276] Pouranvari, M., Ekrami, A., Kokabi, A.H., Materials and Design, 50, 2013, 694-701. https://doi.org/10.1016/j.matdes.2013.03.030

[277] Pouranvari, M., Ekrami, A., Kokabi, A.H., Materiali in Tehnologije, 47[5] 2013, 593-599.

[278] Pouranvari, M., Ekrami, A., Kokabi, A.H., Canadian Metallurgical Quarterly, 53[1] 2014, 38-46. https://doi.org/10.1179/1879139513Y.0000000076

[279] Pouranvari, M., Ekrami, A., Kokabi, A.H., Science and Technology of Welding and Joining, 19[2] 2014, 105-110. https://doi.org/10.1179/1362171813Y.0000000170

[280] Pouranvari, M., Mousavizadeh, S.M., Materiali in Tehnologije, 49[2] 2015, 247-251. https://doi.org/10.17222/mit.2014.048

[281] Pouranvari, M., Ekrami, A., Kokabi, A.H., Journal of Alloys and Compounds, 723, 2017, 84-91. https://doi.org/10.1016/j.jallcom.2017.06.206

[282] Tarai, U.K., Robi, P.S., Pal, S., IOP Conference Series - Materials Science and Engineering, 346[1] 2018, 012048. https://doi.org/10.1088/1757-899X/346/1/012048

[283] Jalilvand, V., Omidvar, H., Shakeri, H.R., Rahimipour, M.R., Materials Characterization, 75, 1970, 20-28. https://doi.org/10.1016/j.matchar.2012.10.004

[284] Su, C.Y., Chou, C.P., Wu, B.C., Lih, W.C., Materials Science and Technology, 15[3] 1999, 316-322. https://doi.org/10.1179/026708399101505743

[285] Idowu, O.A., Richards, N.L., Chaturvedi, M.C., Materials Science and Engineering A, 397[1-2] 2005, 98-112. https://doi.org/10.1016/j.msea.2005.01.055

[286] Ojo, O.A., Abdelfatah, M.M., Materials Science and Technology, 24[6] 2008, 739-743. https://doi.org/10.1179/174328408X281958

[287] Idowu, O.A., Ojo, O.A., Chaturvedi, M.C., Metallurgical and Materials Transactions A, 37[9] 2006, 2787-2796. https://doi.org/10.1007/BF02586111

[288] Ojo, O.A., Richards, N.L., Chaturvedi, M.C., Science and Technology of Welding and Joining, 9[6] 2004, 532-540. https://doi.org/10.1179/174329304X8702

[289] Wikstrom, N.P., Ojo, O.A., Chaturvedi, M.C., Materials Science and Engineering A, 417[1-2] 2006, 299-306. https://doi.org/10.1016/j.msea.2005.10.056

[290] Rie, S., Kazuyoshi, S., Kazutoshi, N., Journal of the Japan Welding Society, 24[1] 2006, 100-107. https://doi.org/10.2207/qjjws.24.100

[291] Mosallaee, M., Ekrami, A., Ohsasa, K., Matsuura, K., Materials Science and Technology, 24[4] 2008, 449-456. https://doi.org/10.1179/174328408X281877

[292] Mosallaee, M., Ekrami, A., Ohsasa, K., Matsuura, K., Metallurgical and Materials Transactions A, 39[10] 2008, 2389-2402. https://doi.org/10.1007/s11661-008-9588-0

[293] Abdelfatah, M.M., Ojo, O.A., Metallurgical and Materials Transactions A, 40[2] 2009, 377-385. https://doi.org/10.1007/s11661-008-9726-8

[294] Jalilvand, V., Omidvar, H., Rahimipour, M.R., Shakeri, H.R., Materials Science and Technology, 29[4] 2013, 439-445. https://doi.org/10.1179/1743284712Y.0000000138

[295] Jalilvand, V., Omidvar, H., Rahimipour, M.R., Shakeri, H.R., Materials and Design, 52, 2013, 36-46. https://doi.org/10.1016/j.matdes.2013.05.042

[296] Binesh, B., Gharehbagh, A.J, Journal of Materials Science and Technology, 32[11] 2016, 1137-1151. https://doi.org/10.1016/j.jmst.2016.07.017

[297] Shakerin, S., Omidvar, H., Mirsalehi, S.E., Materials and Design, 89, 2016, 611-619. https://doi.org/10.1016/j.matdes.2015.10.003

[298] Maleki, V., Omidvar, H., Rahimipour, M.R., Transactions of Nonferrous Metals Society of China, 26[2] 2016, 437-447. https://doi.org/10.1016/S1003-6326(16)64132-8

[299] Adebajo, O.J., Ojo, O.A., Metallurgical and Materials Transactions A, 48[1] 2017, 26-33. https://doi.org/10.1007/s11661-016-3837-4

[300] Payton, L.N., Chitti, S.V., Taarea, D.R., Sofyan, N.I., Gale, W.F., Butts, D.A., Aluru, R., Love, R.D., ASM Conference Proceedings: Joining of Advanced and Specialty Materials, 2004, 67-70.

[301] Yang, Y.H., Xie, Y.J., Wang, M.S., Ye, W., Materials and Design, 51, 2013, 141-147. https://doi.org/10.1016/j.matdes.2013.04.024

[302] Liu, J.D., Li, B., Sun, Y., Jin, T., Sun, X.F., Hu, Z.Q., Rare Metals, 34, 2015, 1-5. https://doi.org/10.1007/s12598-013-0148-4

[303] Ekrami, A., Khan, T.I., Materials Science and Technology, 15[8] 1999, 946-950. https://doi.org/10.1179/026708399101506625

[304] Ekrami, A., Khan, T.I., Malik, H., Materials Science and Technology, 19[1] 2003, 132-136. https://doi.org/10.1179/026708303225008536

[305] Saha, R.K., Khan, T.I., Journal of Materials Engineering and Performance, 15[6] 2006, 722-728. https://doi.org/10.1361/105994906X150786

[306] Saha, R.K., Khan, T.I., Materials Science and Technology Conference and Exhibition, "Exploring Structure, Processing and Applications across Multiple Materials Systems", 5, 2007, 3232-3242.

[307] Nakao, Y., Nishimoto, K., Shinozaki, K., Kang, C.Y., Shigeta, H., Transactions of the Japan Welding Society, 23[2] 1992, 20-25.

[308] Nakao, Y., Shinozaki, K., Nishimoto, K., Kang, C.Y., Shigeta, H., Quarterly Journal of the Japan Welding Society, 9[4] 1991, 550-555. https://doi.org/10.2207/qjjws.9.550

[309] Nakao, Y., Nishimoto, K., Shinozaki, K., Kang, C.Y., Shigeta, H., Quarterly Journal of the Japan Welding Society, 9[4] 1991, 86-91.

[310] Nakao, Y., Nishimoto, K., Shinozaki, K., Kang, Chung Y., Shigeta, H., Transactions of the Japan Welding Society, 23[2] 1992, 26-32.

[311] Nakao, Y., Nishimoto, K., Shinozaki, K., Yun, K.C., Shigeta, H., Quarterly Journal of the Japan Welding Society, 9[1] 1991, 55-62. https://doi.org/10.2207/qjjws.9.55

[312] Nakao, Y., Nishimoto, K., Shinozaki, K., Yun, K.C., Hori, Y., Quarterly Journal of the Japan Welding Society, 6[4] 1988, 519-526. https://doi.org/10.2207/qjjws.6.519

[313] Nakao, Y., Nishimoto, K., Shinozaki, K., Yun, K.C., Quarterly Journal of the Japan Welding Society, 7[2] 1989, 213-219. https://doi.org/10.2207/qjjws.7.213

[314] Nakao, Y., Nishimoto, K., Shinozaki, K., Yun, K.C., Quarterly Journal of the Japan Welding Society, 7[2] 1989, 47-53.

[315] Nakao, Y., Nishimoto, K., Shinozaki, K., Yun, K.C., Hori, Y., Quarterly Journal of the Japan Welding Society, 9[1] 1991, 62-68. https://doi.org/10.2207/qjjws.9.62

[316] Nakao, Y., Nishimoto, K., Shinozaki, K., Kang, C.Y., Shigeta, H., Quarterly Journal of the Japan Welding Society, 9[4] 1991, 92-97.

[317] Chai, L., Huang, J.H., Hou, J.B., Lang, B., Wang, L., Transactions of Materials and Heat Treatment, 35[11] 2014, 117-121.

[318] Lang, B., Hou, J., Guo, D., Chai, L., Transactions of the China Welding Institution, 38[1] 2017, 87-90.

[319] Guo, W., Wang, H., Jia, Q., Peng, P., Zhu, Y., High Temperature Materials and Processes, 36[7] 2017, 677-682. https://doi.org/10.1515/htmp-2015-0243

[320] Chai, L., Huang, J., Hou, J., Lang, B., Wang, L., Journal of Materials Engineering and Performance, 24[6] 2015, 2287-2293. https://doi.org/10.1007/s11665-015-1504-3

[321] Li, W., Jin, T., Sun, X., Guo, Y., Guan, H., Hu, Z., Journal of Materials Science and Technology, 18[1] 2002, 54-56. https://doi.org/10.1016/j.jmst.2015.10.002

[322] Li, W., Jin, T., Hu, Z., Acta Metallurgica Sinica, 44[12] 2008, 1474-1478.

[323] Liu, J.D., Jin, T., Zhao, N.R., Wang, Z.H., Sun, X.F., Guan, H.R., Hu, Z.Q., Materials Characterization, 59[1] 2008, 68-73. https://doi.org/10.1016/j.matchar.2006.10.018

[324] Ekrami, A., Moeinifar, S., Kokabi, A.H., Materials Science and Engineering A, 456[1-2] 2007, 93-98. https://doi.org/10.1016/j.msea.2006.12.044

[325] Yu, Z., Ding, X., Cao, L., Zheng, Y., Feng, Q., Acta Metallurgica Sinica, 52[5] 2016, 549-560.

[326] Wang, J., Shin, S., Hu, A., Wilt, J.K., Computational Materials Science, 152, 2018, 228-235. https://doi.org/10.1016/j.commatsci.2018.05.056

[327] Saijo, S., Koyama, S., Shohji, I., Procedia Engineering, 184, 2017, 284-289. Image reproduced under licence: https://creativecommons.org/licenses/by-nc-nd/4.0/

[328] Brochu, M., Wanjara, P., International Journal of Refractory Metals and Hard Materials, 25[1] 2007, 67-71. https://doi.org/10.1016/j.ijrmhm.2006.01.001

[329] Chen, R.S., Zhang, F.G., Liu, S.C., Liu, D.Y., Transactions of Materials and Heat Treatment, 31[1] 2010, 126-131.

[330] Jing, X.T., Chen, S.J., Lu, J.F., Li, X.G., Transactions of the China Welding Institution, 27[2] 2006, 97-101.

[331] Wang, X., Li, X., Wang, C., Materials Science and Engineering A, 560, 2013, 711-716. https://doi.org/10.1016/j.msea.2012.10.018

[332] Abdolvand, R., Atapour, M., Shamanian, M., Allafchian, A., Journal of Manufacturing Processes, 25, 2017, 172-180. https://doi.org/10.1016/j.jmapro.2016.11.013

[333] Nakahashi, M., Yamazaki, T., Takeda, H., Haga, M., Nippon Kinzoku Gakkaisi, 49[4] 1985, 285-290.

[334] Chen, S., Tang, H., Zhao, P., China Welding, 26[2] 2017, 52-57.

[335] Wang, X.G., LI, X.G., Advanced Materials Research, 712-715, 2013, 701-704.

[336] Chen, S.J., Guo, S.J., Liang, F., Advanced Materials Research, 97-101, 2010, 107-110.

[337] Gao, Z., Chen, S., Xu, Q., Materials Science Forum, 704-705, 2012, 823-827. https://doi.org/10.4028/www.scientific.net/MSF.704-705.823

[338] Bigvand, A.G., Ojo, O.A., Metallurgical and Materials Transactions A, 45[4] 2014, 1670-1674. https://doi.org/10.1007/s11661-014-2208-2

[339] Liu, J., Cao, J., Lin, X., Song, X., Feng, J., Materials and Design, 49, 2013, 622-626. https://doi.org/10.1016/j.matdes.2013.02.022

[340] Steuer, S., Piegert, S., Frommherz, M., Singer, R.F., Scholz, A., Advanced Materials Research, 278, 2011, 454-459.

[341] Steuer, S., Singer, R.F., Metallurgical and Materials Transactions A, 44[5] 2013, 2226-2232. https://doi.org/10.1007/s11661-012-1597-3

[342] Steuer, S., Singer, R.F., Metallurgical and Materials Transactions A, 45[8] 2014, 3545-3553. https://doi.org/10.1007/s11661-014-2304-3

[343] Aluru, R., Sofyan, N.I., Fergus, J.W., Gale, W.F., ASM Proceedings of the International Conference: Trends in Welding Research, 2005, 879-883.

[344] Aluru, R., Gale, W.F., Chitti, S.V., Sofyan, N., Love, R.D., Fergus, J.W., Materials Science and Technology, 24[5] 2008, 517-528. https://doi.org/10.1179/174328408X293478

[345] Aluru, R., Sofyan, N.I., Gale, W.F., Transactions of the Indian Institute of Metals, 59[2] 2006, 185-191.

[346] Liu, J.D., Jin, T., Zhao, N.R., Wang, Z.H., Sun, X.F., Guan, H.R., Hu, Z.Q., Science and Technology of Welding and Joining, 15[3] 2010, 194-198. https://doi.org/10.1179/136217109X12518083193513

[347] Liu, J.D., Jin, T., Zhao, N.R., Wang, Z.H., Sun, X.F., Guan, H.R., Hu, Z.Q., Materials Characterization, 62[5] 2011, 545-553. https://doi.org/10.1016/j.matchar.2011.03.012

[348] Arafin, M.A., Medraj, M., Turner, D.P., Bocher, P., Advanced Materials Research, 15-17, 2007, 882-887.

[349] Arafin, M.A., Medraj, M., Turner, D.P., Bocher, P., Materials Science and Engineering A, 447[1-2] 2007, 125-133. https://doi.org/10.1016/j.msea.2006.10.045

[350] Wu, X., Chandel, R.S., Li, H., Journal of Materials Science, 36[6] 2001, 1539-1546. https://doi.org/10.1023/A:1017513200502

[351] Wikstrom, N.P., Idowu, O.A., Ojo, O.A., Chaturvedi, M.C., Proceedings of the 3rd International Brazing and Soldering Conference, 2006, 6-11.

[352] Khakian, M., Nategh, S., Mirdamadi, S., Journal of Alloys and Compounds, 653, 2015, 386-394. https://doi.org/10.1016/j.jallcom.2015.09.044

[353] Ghomi, M.K., Nategh, S., Mirdamadi, S., Materiali in Tehnologije, 50[3] 2016, 365-371. https://doi.org/10.17222/mit.2015.072

[354] Gale, W.F., Orel, S.V., Metallurgical and Materials Transactions A, 27[7] 1996, 1925-1931. https://doi.org/10.1007/BF02651942

[355] Kitchings, M.K., Guan, Y., Gale, W.F., Microstructural Science, 26, 1998, 425-430.

[356] Gale, W.F., Guan, Y., Materials Science and Technology, 15[4] 1999, 464-467. https://doi.org/10.1179/026708399101505950

[357] Guan, Y., Gale, W.F., Materials Science and Technology, 15[2] 1999, 207-212. https://doi.org/10.1179/026708399101505608

[358] Gale, W.F., Guan, Y., Orel, S.V., International Journal of Materials and Product Technology, 13[1-2] 1998, 1-12.

[359] Abdo, Z.A.M., Guan, Y., Gale, W.F., Materials Research Society Symposium - Proceedings, 552, 1999, KK9.4.1-KK9.4.7.

[360] Gale, W.F., Guan, Y., ASM Proceedings of the International Conference: Trends in Welding Research, 1998, 663-668.

[361] Gale, W.F., Guan, Y., Journal of Materials Science, 34[5] 1999, 1061-1071. https://doi.org/10.1023/A:1004504330294

[362] Gale, W.F., Wen, X., Materials Science and Technology, 17[4] 2001, 459-464. https://doi.org/10.1179/026708301101510050

[363] Gale, W.F., Guan, Y., Journal of Materials Science, 32[2] 1997, 357-364. https://doi.org/10.1023/A:1018597115287

[364] Gale, W.F., Guam, Y., Journal of Materials Science, 32[10] 1997, 2543-2547. https://doi.org/10.1023/A:1018590115055

[365] Rahman, A.H.M.E., Abu-Mahfouz, I., Materials Science and Technology Conference and Exhibition, 2, 2016, 1089-1096.

[366] Ren, H.S., Xiong, H.P., Pang, S.J., Chen, B., Wu, X., Cheng, Y.Y., Chen, B.Q., Metallurgical and Materials Transactions A, 47[4] 2016, 1668-1676. https://doi.org/10.1007/s11661-015-3310-9

[367] Ren, H., Xiong, H., Chen, B., Pang, S., Ye, L., Transactions of the China Welding Institution, 37[3] 2016, 106-110.

[368] Ren, H.S., Xiong, H.P., Chen, B., Pang, S.J., Wu, X., Cheng, Y.Y., Chen, B.Q., Materials Science and Engineering A, 651, 2016, 45-54. https://doi.org/10.1016/j.msea.2015.10.075

[369] Duan, H., Koçak, M., Bohm, K.H., Ventzke, V., Science and Technology of Welding and Joining, 9[6] 2004, 525-531. https://doi.org/10.1179/136217104225021869

[370] Duan, H.P., Bohm, K.H., Ventzke, V., Kocak, M., Transactions of Nonferrous Metals Society of China, 15[2] 2005, 375-378.

[371] Emadinia, O., Simões, S., Viana, F., Vieira, M.F., Cavaleiro, A.J., Ramos, A.S., Vieira, M.T., Welding in the World, 60[2] 2016, 337-344. https://doi.org/10.1007/s40194-015-0282-8

[372] Simões, S., Viana, F., Ramos, A.S., Vieira, M.T., Vieira, M.F., Journal of Materials Science, 48[21] 2013, 7718-7727. https://doi.org/10.1007/s10853-013-7592-2

[373] Gu, X.Y., Liu, Y.J., Sun, D.Q., Xu, F., Meng, L.S., Gao, S., Journal of Jilin University - Engineering and Technology, 47[5] 2017, 1534-1541.

[374] Araki, T., Koba, M., Nambu, S., Inoue, J., Koseki, T., Materials Transactions, 52[3] 2011, 568-571. https://doi.org/10.2320/matertrans.M2010359

[375] Zhang, G., Zhang, J., Zhong, M., Pei, Y., Transactions of the China Welding Institution, 28[9] 2007, 59-62.

[376] Du, S.M., Qin, Q., Advanced Materials Research, 937, 2014, 172-177.

[377] Liu, M., Sheng, G., Transactions of the China Welding Institution, 35[5] 2014, 39-42.

[378] Liu, M.E., Sheng, G.M., Yin, L.J., Journal of Functional Materials, 43[17] 2012, 2401-2403+2407.

[379] Liu, M.E., Sheng, G.M., Journal of Central South University - Science and Technology, 43[7] 2012, 2542-2546.

[380] Du, S.M., Liu, G., Wang, M.J., Chinese Journal of Nonferrous Metals, 23[5] 2013, 1255-1261.

[381] Du, S.M., Zhang, Y.Q., Du, C., Hu, J., IOP Conference Series: Materials Science and Engineering, 170[1] 2017, 012014. https://doi.org/10.1088/1757-899X/170/1/012014

[382] Atabaki, M.M., Idris, J., Mullis, A., Proceedings of the 6th International Quenching and Control of Distortion Conference, 2012, 93-103.

[383] Atabaki, M.M., Idris, J., Materials and Design, 34, 2012, 832-841. https://doi.org/10.1016/j.matdes.2011.07.021

[384] Samavatian, M., Khodabandeh, A., Halvaee, A., Amadeh, A.A., Transactions of Nonferrous Metals Society of China, 25[3] 2015, 63662, 770-775.

[385] Anbarzadeh, A., Sabet, H., Abbasi, M., Materials Letters, 178, 2016, 280-283. https://doi.org/10.1016/j.matlet.2016.04.071

[386] Samavatian, M., Halvaee, A., Amadeh, A., Zakipour, S., Journal of Materials Engineering and Performance, 24[6] 2015, 2526-2534. https://doi.org/10.1007/s11665-015-1512-3

[387] AlHazaa, A., Khan, T.I., Haq, I., Materials Characterization, 61[3] 2010, 312-317. https://doi.org/10.1016/j.matchar.2009.12.014

[388] Kenevisi, M.S., Mousavi Khoie, S.M., Materials and Design, 38, 2012, 19-25. https://doi.org/10.1016/j.matdes.2012.01.046

[389] Hadibeyk, S., Beidokhti, B., Sajjadi, S.A., Journal of Alloys and Compounds, 731, 2018, 929-935. https://doi.org/10.1016/j.jallcom.2017.10.105

[390] Hadibeyk, S., Beidokhti, B., Sajjadi, S.A., Journal of Materials Processing Technology, 255, 2018, 673-678. https://doi.org/10.1016/j.jmatprotec.2018.01.022

[391] Ojo, O.A., Aina, O., Metallurgical and Materials Transactions A, 49[5] 2018, 1481-1485.

[392] Ojo, O.A., Olatunji, O.A., Chaturvedi, M.C., Philosophical Magazine Letters, 97[11] 2017, 419-428. https://doi.org/10.1080/09500839.2017.1396373

[393] Xiong, J.T., Xie, Q., Li, J.L., Zhang, F.S., Huang, W.D., Journal of Materials Engineering and Performance, 21[1] 2012, 33-37. https://doi.org/10.1007/s11665-011-9870-y

[394] Kaya, Y., Kahraman, N., Durgutlu, A., Gülenç, B., Proceedings of the Institution of Mechanical Engineers B, 226[3] 2012, 478-484. https://doi.org/10.1177/0954405411423333

[395] Liu, S.C., Chen, R.S., Liu, D.Y., Transactions of the China Welding Institution, 28[1] 2007, 21-24.

[396] Yu, Z., Wang, M., Wang, F., Wang, Y., Qi, K., Transactions of the China Welding Institution, 21[3] 2000, 32-35.

[397] Yu, Z.S., Wang, F.J., Li, X.Q., Wang, Y., Wu, M.F., Transactions of Nonferrous Metals Society of China, 10[3] 2000, 349-352.

[398] Shen, Q., Xiang, H., Luo, G., Wang, C., Li, M., Zhang, L., Materials Science and Engineering A, 596, 2014, 45-51. https://doi.org/10.1016/j.msea.2013.12.017

[399] Shen, Y.F., Liu, S.F., Li, J.W., Xu, J., Transactions of the China Welding Institution, 26[12] 2005, 73-76.

[400] Peng, J., Wang, R.C., Liu, H.S., Li, J.Y., Journal of Materials Science - Materials in Electronics, 29[1] 2018, 313-322. https://doi.org/10.1007/s10854-017-7918-y

[401] Du, S., Gao, Y., Hu, J., Rare Metal Materials and Engineering, 45[8] 2016, 2064-2070.

[402] Konieczny, M., Szwed, B., Mola, R., METAL 2015, - 24th International Conference on Metallurgy and Materials, Conference Proceedings, 2015, 1513-1518.

[403] Wang, Y.L., Li, H., Li, Z.X., Feng, J.C., Journal of Materials Engineering, 9, 2008, 48-55.

[404] Li, H., Li, Z.X., Wang, Y.L., Feng, J.C., Journal of Beijing University of Technology, 35[S] 2009, 88-92.

[405] Dong, H., Yang, Z., Wang, Z., Deng, D., Dong, C., Journal of Materials Engineering and Performance, 23[10] 2014, 3770-3777. https://doi.org/10.1007/s11665-014-1145-y

[406] Chen, H., Long, C., Wei, T., Gao, W., Xiao, H., Chen, L., Materials and Design, 60, 2014, 358-362. https://doi.org/10.1016/j.matdes.2014.03.055

[407] Zhang, L.X., Sun, Z., Xue, Q., Lei, M., Tian, X.Y., Materials and Design, 90, 2016, 949-957. https://doi.org/10.1016/j.matdes.2015.11.041

[408] Elrefaey, A., Tillmann, W., Advanced Engineering Materials, 11[7] 2009, 556-560. https://doi.org/10.1002/adem.200900021

[409] Elrefaey, A., Tillmann, W., Journal of Materials Processing Technology, 209[5] 2009, 2746-2752. https://doi.org/10.1016/j.jmatprotec.2008.06.014

[410] Wang, Y.L., Li, H., Li, Z.X., Feng, J.C., Journal of Materials Engineering, 9, 2008, 48-51+55.

[411] Wang, Y., Li, H., Li, Z., Feng, J., Transactions of the China Welding Institution, 30[4] 2009, 77-80.

[412] Li, H., Li, Z., Wang, Y., Feng, J., Rare Metal Materials and Engineering, 40[8] 2011, 1382-1386.

[413] Wang, Y., Li, H., Li, Z., Feng, J., Transactions of the China Welding Institution, 30[4] 2009, 77-80.

[414] Wang, Y.L., Gao, Q.Z., Sun, G.F., Ye, J., Advanced Materials Research, 750-752, 2013, 739-742.

[415] Balasubramanian, M., Materials and Design, 77, 2015, 161-169. https://doi.org/10.1016/j.matdes.2015.04.003

[416] Norouzi, E., Atapour, M., Shamanian, M., Journal of Alloys and Compounds, 701, 2017, 335-341. https://doi.org/10.1016/j.jallcom.2017.01.091

[417] Norouzi, E., Atapour, M., Shamanian, M., Allafchian, A., Materials and Design, 99, 2016, 543-551. https://doi.org/10.1016/j.matdes.2016.03.101

[418] Simões, S., Viana, F., Ramos, A.S., Vieira, M.T., Vieira, M.F., Materials Chemistry and Physics, 171, 2016, 73-82. https://doi.org/10.1016/j.matchemphys.2015.11.032

[419] Simões, S., Ramos, A.S., Viana, F., Vieira, M.T., Vieira, M.F., Metals, 6[5] 2016, 96 https://doi.org/10.3390/met6050096

[420] Zakipour, S., Halvaee, A., Amadeh, A.A., Samavatian, M., Khodabandeh, A., Journal of Alloys and Compounds, 626, 2015, 269-276. https://doi.org/10.1016/j.jallcom.2014.11.160

[421] Jalali, A., Atapour, M., Shamanian, M., Vahman, M., Journal of Manufacturing Processes, 33, 2018, 194-202. https://doi.org/10.1016/j.jmapro.2018.05.014

[422] Kundu, S., Thirunavukarasu, G., Chatterjee, S., Mishra, B., Metallurgical and Materials Transactions A, 46[12] 2015, 5756-5771. https://doi.org/10.1007/s11661-015-3142-7

[423] Szwed, B., Konieczny, M., Mola, R., 24th International Conference on Metallurgy and Materials, Conference Proceedings, 2015, 1662-1667.

[424] Szwed, B., Konieczny, M., METAL 2016, - 25th Anniversary International Conference on Metallurgy and Materials, Conference Proceedings, 2016, 1552-1557.

[425] Ma, Y., Zhu, W., Cai, Q., Liu, W., Pang, X., International Journal of Refractory Metals and Hard Materials, 73, 2018, 91-98. https://doi.org/10.1016/j.ijrmhm.2018.02.002

[426] Ma, Y., Wang, Y., Liu, W., Cai, Q., Transactions of the China Welding Institution, 34[12] 2013, 17-20.

[427] Zhong, Z., Hinoki, T., Nozawa, T., Park, Y.H., Kohyama, A., Journal of Alloys and Compounds, 489[2] 2010, 545-551. https://doi.org/10.1016/j.jallcom.2009.09.105

[428] Yang, Z.H., Shen, Y.F., Li, X.Q., Meng, Q.B., Chinese Journal of Nonferrous Metals, 22[10] 2012, 2783-2789.

[429] Atabaki, M.M., Hanzaei, A.T., Materials Characterization, 61[10] 2010, 982-991. https://doi.org/10.1016/j.matchar.2010.06.010

[430] Atabaki, M.M., Idris, J., Journal of Manufacturing Science and Engineering, Transactions of the ASME, 134[1] 2012, 015001.

[431] Atabaki, M.M., Journal of Nuclear Materials, 406[3] 2010, 330-344. https://doi.org/10.1016/j.jnucmat.2010.09.003

[432] Atabaki, M.N., Bajgholi, M.E., Dehkordi, E.H., Materials and Design, 42, 2012, 172-183. https://doi.org/10.1016/j.matdes.2012.05.040

[433] Xiong, J., Zhang, F., Li, J., Huang, W., Rare Metal Materials and Engineering, 35[10] 2006, 1677-1680.

[434] Atieh, A.M., Khan, T.I., Journal of Materials Science, 49[22] 2014, 7648-7658. https://doi.org/10.1007/s10853-014-8473-z

[435] Atieh, A.M., Khan, T.I., Science and Technology of Welding and Joining, 19[4] 2014, 333-342. https://doi.org/10.1179/1362171814Y.0000000196

[436] Atieh, A.M., Khan, T.I., IOP Conference Series - Materials Science and Engineering, 60, 2014, 012036. Image reproduced under licence: Creative Commons Attribution 3.0. doi:10.1088/1757-899X/60/1/012036 https://doi.org/10.1088/1757-899X/60/1/012036

[437] Gu, X.Y., Sun, D.Q., Liu, L., Duan, Z.Z., Journal of Alloys and Compounds, 476[1-2] 2009, 492-499. https://doi.org/10.1016/j.jallcom.2008.09.021

[438] Duan, H., Luo, J., Zhang, T., Kocak, M., Journal of Beijing University of Aeronautics and Astronautics, 30[10] 2004, 984-988.

[439] Ren, H., Xiong, H., Wu, X., Chen, B., Cheng, Y., Chen, B., Journal of Mechanical Engineering, 53[4] 2017, 1-10. https://doi.org/10.3901/JME.2017.04.001

[440] Wang, Y.H., Qi, X.S., Meng, X.L., Li, W.B., Wang, C.Y., Kou, H.C., Li, J.S., Advanced Materials Research, 753-755, 2013, 396-401.

[441] Tang, B., Qi, X.S., Kou, H.C., Li, J.S., Milenkovic, S., Advanced Engineering Materials, 18[4] 2016, 657-664. https://doi.org/10.1002/adem.201500457

[442] Ren, H.S., Wu, X., Chen, B., Xiong, H.P., Cheng, Y.Y., Welding in the World, 61[2] 2017, 375-381. https://doi.org/10.1007/s40194-016-0415-8

[443] Qian, J.W., Li, J.L., Hou, J.B., Xiong, J.T., Zhang, F.S., Han, Z.C., Journal of Aeronautical Materials, 29[1] 2009, 57-62.

[444] Liu, K., Li, Y., Xia, C., Wang, J., Vacuum, 143, 2017, 195-198. https://doi.org/10.1016/j.vacuum.2017.06.025

[445] Ghoneim, A., Ojo, O.A., Metallurgical and Materials Transactions A, 43[3] 2012, 900-911. https://doi.org/10.1007/s11661-011-1010-7

[446] Cook, G.O., Sorensen, C.D., Metallurgical and Materials Transactions A, 44[13] 2013, 5732-5753. https://doi.org/10.1007/s11661-013-1956-8

[447] Cook, G.O., Sorensen, C.D., Metallurgical and Materials Transactions A, 44[13] 2013, 5732-5753 and Metallurgical and Materials Transactions A, 44[13] 2013, 5754-5772.

[448] Wu, M., Kuang, H., Wang, F., Lin, H., Xu, G., Acta Metallurgica Sinica, 50[5] 2014, 619-625.

[449] Lang, F., Yamaguchi, H., Nakagawa, H., Sato, H., Journal of the Electrochemical Society, 160[8] 2013, D315-D319. https://doi.org/10.1149/2.114308jes

[450] Hynes, N.R.J., Raja, M.K., AIP Conference Proceedings, 1728, 2016, 020543.

[451] Li, J.L., Xiong, J.T., Zhang, F.S., Materials Science and Engineering A, 483-484[1-2] 2008, 698-700. https://doi.org/10.1016/j.msea.2006.11.170

[452] Zhang, X., Shi, X., Wang, J., Li, H., Li, K., Ren, Y., Acta Metallurgica Sinica, 27[4] 2014, 663-669.

[453] Wang, W., Fan, D., Huang, J., Cui, B., Chen, S., Zhao, X., Materials Letters, 143, 2015, 237-240. https://doi.org/10.1016/j.matlet.2014.12.110

[454] Fan, D., Li, C., Huang, J., Yang, J., Cui, B., Wang, W., Ceramics International, 43[15] 2017, 13009-13012. https://doi.org/10.1016/j.ceramint.2017.06.044

[455] Hynes, N.R.J., Velu, P.S., Kumar, R., Raja, M.K., Ceramics International, 43[10] 2017, 7762-7767. https://doi.org/10.1016/j.ceramint.2017.03.084

[456] Blue, C.A., Sikka, V.K., Blue, R.A., Lin, R.Y., Metallurgical and Materials Transactions A, 27[12] 1996, 4011-4018. https://doi.org/10.1007/BF02595650

[457] Huang, L., Sheng, G.M., Li, J., Huang, G.J., Yuan, X.J., Journal of Central South University, 25[5] 2018, 1025-1032. https://doi.org/10.1007/s11771-018-3802-z

[458] Li, J., Sheng, G.M., Journal of Materials Engineering, [12] 2014, 60-65.

[459] Kliauga, A.M., Travessa, D., Ferrante, M., Materials Characterization, 46[1] 2001, 65-74. https://doi.org/10.1016/S1044-5803(00)00095-4

[460] Zhou, S., Li, X., Xiong, W., Zhou, Y., Journal of Wuhan University of Technology - Materials Science, 24[3] 2009, 432-439.

[461] Guo, Y., Gao, B., Liu, G., Zhou, T., Qiao, G., International Journal of Refractory Metals and Hard Materials, 51, 2015, 250-257. https://doi.org/10.1016/j.ijrmhm.2015.04.018

[462] Kuromitsu, Y., Nagatomo, Y., Akiyama, K., Shibata, N., Ikuhara, Y., Journal of the Ceramic Society of Japan, 125[3], 2017, 165-167. https://doi.org/10.2109/jcersj2.16235

[463] Dezellus, O., Andrieux, J., Bosselet, F., Sacerdote-Peronnet, M., Baffie, T., Hodaj, F., Eustathopoulos, N., Viala, J.C., Materials Science and Engineering A, 495[1-2] 2008, 254-258. https://doi.org/10.1016/j.msea.2007.10.104

[464] Lan, L., Xuan, W., Wang, J., Li, C., Ren, Z., Yu, J., Peng, J., Vacuum, 130, 2016, 105-108. https://doi.org/10.1016/j.vacuum.2016.04.033

[465] Lan, L., Yu, J., Yang, Z., Li, C., Ren, Z., Wang, Q., Ceramics International, 42[1] 2016, 1633-1639. https://doi.org/10.1016/j.ceramint.2015.09.115

[466] Lan, L., Ren, Z., Yu, J., Yang, Z., Zhong, Y., Materials Letters, 121, 2014, 223-226. https://doi.org/10.1016/j.matlet.2014.01.072

[467] Brochu, M., Pugh, M., Drew, R.A.L., ASM Conference Proceedings - Joining of Advanced and Specialty Materials, 2001, 44-50.

[468] Khan, T.I., Roy, B.N., Journal of Materials Science, 39[2] 2004, 741-743. https://doi.org/10.1023/B:JMSC.0000011546.44307.42

[469] Kim, J.J., Park, J.W., Eagar, T.W., Materials Science and Engineering A, 344[1-2] 2003, 240-244. https://doi.org/10.1016/S0921-5093(02)00402-1

[470] Zhai, Y., North, T.H., Serrato-Rodrigues, J., Journal of Materials Science, 32[6] 1997, 1393-1397. https://doi.org/10.1023/A:1018529228375

[471] Zhai, Y., North, T.H., Journal of Materials Science, 32[21] 1997, 5571-5575. https://doi.org/10.1023/A:1018624507922

[472] Zhang, C., Qiao, G., Jin, Z., Journal of the European Ceramic Society, 22[13] 2002,

2181-2186 and Rare Metal Materials and Engineering, 31[4] 2002, 302.

[473] Chunguang, Z., Guanjun, Q., Zhihao, J., Rare Metal Materials and Engineering, 31[4] 2002, 302.

[474] Zhai, Y., North, T.H., Ren, J., Journal of Materials Science, 32[6] 1997, 1399-1404. https://doi.org/10.1023/A:1018581312445

[475] Zhang, J.X, Chandel, R.S, Chen, Y.Z, Seow, H.P., Journal of Materials Processing Technology, 122[2-3] 2002, 220-225. https://doi.org/10.1016/S0924-0136(02)00010-9

[476] Jiang, X., Li, J., Liu, W., Zhu, D., Wang, B., Asian Journal of Chemistry, 26[17] 2014, 5682-5686. https://doi.org/10.14233/ajchem.2014.18188

[477] Guo, Y., Wang, Y., Gao, B., Shi, Z., Yuan, Z., Ceramics International, 42[15] 2016, 16729-16737. https://doi.org/10.1016/j.ceramint.2016.07.145

[478] Zhang, J.X., Chandel, R.S., Seow, H.P., ASM Conference Proceedings: Joining of Advanced and Specialty Materials, 2001, 56-60.

[479] Zhao, Q.Z., Chen, Z., Zhou, J.S., Journal of East China Shipbuilding Institute, 15[2] 2001, 8-11.

[480] Yi, J., Zhang, Y., Hu, H., Wang, X., Chen, H., Dai, M., Rare Metal Materials and Engineering, 43[11] 2014, 2593-2596. https://doi.org/10.1016/S1875-5372(15)60020-0

[481] Long, Z., Dai, B., Tan, S., Wang, Y., Wei, X., Ceramics International, 43[18] 2017, 17000-17004. https://doi.org/10.1016/j.ceramint.2017.09.108

[482] Yu, Z.S., Liang, C., Li, R.F., Wu, M.F., Qi, K., Transactions of Nonferrous Metals Society of China, 14[1] 2004, 99-104.

[483] Tillmann, W., Schaak, C., Pfeiffer, J., Materials Science and Technology Conference and Exhibition 2015, MS and T 2015, 1, 2015, 403-410.

Materials Research Foundations **43** (2019)   doi: http://dx.doi.org/10.21741/ 9781644900055

Keyword Index

www.ingramcontent.com/pod-product-compliance
Lightning Source LLC
Chambersburg PA
CBHW070728220326
41598CB00024BA/3353